# DECOMMISSIONING OF NUCLEAR FACILITIES

## FEASIBILITY, NEEDS AND COSTS

Report by an Expert Group

**NUCLEAR ENERGY AGENCY**
ORGANISATION FOR ECONOMIC CO-OPERATION AND DEVELOPMENT

Pursuant to article 1 of the Convention signed in Paris on 14th December, 1960, and which came into force on 30th September, 1961, the Organisation for Economic Co-operation and Development (OECD) shall promote policies designed:

- to achieve the highest sustainable economic growth and employment and a rising standard of living in Member countries, while maintaining financial stability, and thus to contribute to the development of the world economy;
- to contribute to sound economic expansion in Member as well as non-member countries in the process of economic development; and
- to contribute to the expansion of world trade on a multilateral, non-discriminatory basis in accordance with international obligations.

The Signatories of the Convention on the OECD are Austria, Belgium, Canada, Denmark, France, the Federal Republic of Germany, Greece, Iceland, Ireland, Italy, Luxembourg, the Netherlands, Norway, Portugal, Spain, Sweden, Switzerland, Turkey, the United Kingdom and the United States. The following countries acceded subsequently to this Convention (the dates are those on which the instruments of accession were deposited): Japan (28th April, 1964), Finland (28th January, 1969), Australia (7th June, 1971) and New Zealand (29th May, 1973).

The Socialist Federal Republic of Yugoslavia takes part in certain work of the OECD (agreement of 28th October, 1961).

*The OECD Nuclear Energy Agency (NEA) was established on 20th April, 1972, replacing OECD's European Nuclear Energy Agency (ENEA) on the adhesion of Japan as a full Member.*

*NEA now groups all the European Member countries of OECD and Australia, Canada, Japan, and the United States. The Commission of the European Communities takes part in the work of the Agency.*

*The primary objectives of NEA are to promote co-operation between its Member governments on the safety and regulatory aspects of nuclear development, and on assessing the future role of nuclear energy as a contributor to economic progress.*

*This is achieved by:*

- *encouraging harmonisation of governments' regulatory policies and practices in the nuclear field, with particular reference to the safety of nuclear installations, protection of man against ionising radiation and preservation of the environment, radioactive waste management, and nuclear third party liability and insurance;*
- *keeping under review the technical and economic characteristics of nuclear power growth and of the nuclear fuel cycle, and assessing demand and supply for the different phases of the nuclear fuel cycle and the potential future contribution of nuclear power to overall energy demand;*
- *developing exchanges of scientific and technical information on nuclear energy, particularly through participation in common services;*
- *setting up international research and development programmes and undertakings jointly organised and operated by OECD countries.*

*In these and related tasks, NEA works in close collaboration with the International Atomic Energy Agency in Vienna, with which it has concluded a Co-operation Agreement, as well as with other international organisations in the nuclear field.*

Publié en français sous le titre:

DÉCLASSEMENT
DES INSTALLATIONS NUCLÉAIRES
FAISABILITÉ, BESOINS ET COÛTS

# FOREWORD

This study provides an assessment of the feasibility and techniques for decommissioning currently operating large, commercial, nuclear facilities. It also includes estimates of some of the costs of decommissioning.

The report is intended to assist national authorities in Member countries in making decisions concerning decommissioning strategies as well as to inform a wider public of the current state of the art.

The work has been carried out by an international group of experts under the auspices of the NEA Committee for Technical and Economic Studies on Nuclear Energy Development and the Fuel Cycle (FCC). The report does not necessarily represent the views of Member governments or participating organisations.

# TABLE OF CONTENTS

# EXECUTIVE SUMMARY

**General**

The useful operating lifetime of nuclear facilities is largely determined by economic considerations. By some refurbishment and upgrading the operations can probably be extended well beyond the design lifetime, but at some point in time it becomes technically and economically advantageous to replace the facility. Decommissioning means all those activities that begin after operations have ceased with the intention of placing the facility in a condition that provides protection for the health and safety of the decommissioning worker, the public, and the environment.

While the number of reactors likely to be decommissioned within the next ten years is fairly small, a growing number will reach the age of several decades by early in the next century. If the lifetime of all reactors is assumed conservatively to be 25 years, almost three hundred reactors would have to be decommissioned by 2010. In reality, considerably longer lifetimes are expected for many current reactors, and so this number is likely to be an overestimate.

There are three major alternatives or stages of decommissioning. Stage 1 decommissioning consists of minimal decontamination, draining of liquid systems, disconnection of operating systems, physical and administrative controls to assure limited access, and continued surveillance and maintenance for a predetermined time period. Before completing Stage 1 for nuclear reactors, spent fuel has to be removed from the facility. In Stage 2 decommissioning, all equipment and buildings which can be easily dismantled are removed or are decontaminated and made available for other use. Any remaining liquids are drained from the systems. At power plants the biological shield is extended and sealed to completely enclose the reactor structure. At fuel cycle facilities, the primary radioactive plant and equipment will sometimes be removed. Surveillance around the barrier may be reduced, but it is desirable to continue periodic spot checks as well as surveillance of the environment. Stage 3 decommissioning involves the decontamination of materials, equipment and buildings or the removal of these items if decontamination to a specified activity limit is not practicable. The buildings may be demolished and the site made available for re-use, although after decontamination some of the buildings themselves may be re-used.

A number of factors have to be considered when selecting the optimum strategy for the decommissioning of a nuclear facility. These include the national nuclear strategy, characteristics of the facility, health and safety, environmental protection, radioactive waste management, future use of the site, improvement of decommissioning technology that may be achieved in the future, cost and availability of funds for the project and various social considerations. The relative importance of these factors has to be judged case by case.

**Feasibility and development needs**

All three stages of decommissioning have been performed for small test, training, and power demonstration reactors and supporting fuel cycle facilities. Several major projects are under way or are being planned in OECD countries to demonstrate the decommissioning of larger reactors and fuel cycle facilities. Experience from decommissioning nuclear reactors already covers major reactor types and includes the isolation of systems, handling of toxic as well as radioactive material, use of controlled explosives for pipecutting and concrete demolition, use of various decontamination methods, remote segmentation of the pressure vessel and internals, and surveillance practices. Some significant experience has been obtained in the decommissioning of fuel cycle facilities including decontamination, sectioning and removal of glove boxes and equipment inside these glove boxes such as storage tanks and furnaces. In several cases buildings containing fuel fabrication facilities have been decontaminated and re-used. A highly radioactive fuel reprocessing plant has been decontaminated successfully for possible re-use.

Present technology has proved to be quite satisfactory for decommissioning of nuclear facilities to any of the three stages. Though the experience to date is restricted to small facilities, the same technology is applicable to decommissioning of larger commercial-size facilities. Since the present-day reactors will have been operated for longer periods and at higher power levels than the reactors so far decommissioned, some adjustments in the working procedures are necessary to allow for the higher radiation levels; but the radiation from components is not expected to necessitate an essentially new approach. For instance, the reactor vessel and internals must be disassembled by remote operations in both small and large reactors.

Future decommissioning projects will also benefit from the experience obtained from ongoing maintenance and repair work on operating commercial-scale reactors and fuel cycle facilities. Major reactor repair operations with applications to decommissioning include decontamination of primary loop components, remote repair of small piping, fuel-channel pressure-tube replacement, replacement of calandrias, and replacement of steam generators. Experience has also been obtained from post-accident decontamination and recovery. For fuel reprocessing plants and plutonium fabrication facilities most repair work is done remotely, whereas in other fuel cycle facilities direct access is feasible with emphasis on the control of airborne contamination. All these operations have provided information and experience on working safely and cost-effectively under restrictive conditions.

Although current technology is sufficient for decommissioning of commercial-scale facilities, continued development in some areas is desirable. Such areas include decontamination methods, remotely operated equipment for facility and plant equipment disassembly, techniques for minimising waste generation through treatment, waste volume reduction, and discrimination of radioactivity levels in waste.

Furthermore, facilities will be needed for disposal of decommissioning wastes. The wastes from reactor decommissioning are primarily low-level wastes and can be disposed of in the same or similar facilities as the wastes from reactor operations. The volume of radioactive decommissioning wastes from one reactor is of the same order of magnitude as the volume of wastes from its lifetime operations. Wide experience already exists on the handling of reactor wastes and disposal facilities for such wastes are being developed in many countries. The disposal of reactor decommissioning wastes will not require new technical approaches.

Some of the decommissioning wastes from fuel cycle facilities will be high-level or transuranic waste and will require appropriate disposal facilities. However, the facilities being developed for spent fuel and reprocessing wastes can easily handle the small volumes of these decommissioning wastes as well, so that no special facilities will be required. In general, fuel cycle facilities are relatively few and the small decommissioning waste volumes they generate will not make a significant contribution to the total waste volumes.

## Costs

The total costs of decommissioning are dependent on the sequence and timing of the stages that comprise the total decommissioning program. Deferment of an individual stage tends to reduce its costs, but the effect is compromised by increasing storage costs. Any cost comparison of decommissioning strategies is heavily influenced by the discount rate used.

Cost estimates for decommissioning nuclear power plants have been made by several countries. Modified to correspond to a common plant size assumption, they indicate a range from about $95 to some $120 million (January 1984 U.S. dollars) for immediate dismantling of a 1 300 MW(e) pressurised light water reactor (PWR), and some $125 to $175 million for a similar size boiling water reactor (BWR) or pressurised heavy water reactor (PHWR).

According to the estimated undiscounted costs for PWRs and BWRs the deferment of Stage 3 by 30 years after Stage 1 would somewhat increase the costs in relation to immediate Stage 3. According to the estimate made for PHWRs the effect would be opposite. However, the introduction of discounting by 5 per cent per annum would make delayed dismantling the preferred alternative in all cases studied (see Table S1).

Table S-1

**ESTIMATED TOTAL COSTS FOR VARIOUS DECOMMISSIONING STRATEGIES**[a]

Millions of January 1984 U.S. dollars

|  | Canada | Federal Republic of Germany | | Finland | | Sweden | | United States | |
|---|---|---|---|---|---|---|---|---|---|
|  | HWR | PWR | BWR | PWR | BWR | PWR | BWR | PWR | BWR |
| Undiscounted costs: | | | | | | | | | |
| Stage 3 immediately | 145 | 119 | 173 | 105 | – | 107 | 140 | 97 | 113 |
| Stage 1/30 years storage/ Stage 3 | 117 | 121 | 181 | – | 126 | – | – | 121 | 141 |
| Stage 2/100 years storage/ Stage 3 | – | – | – | – | – | – | – | 158 | 186 |
| Costs discounted at 5 % to the year of shutdown: | | | | | | | | | |
| Stage 3 immediately | 129 | 105 | 153 | 93 | – | 95 | 124 | 86 | 100 |
| Stage 1/30 years storage/ Stage 3 | 29 | 30 | 44 | – | 29 | – | – | 41 | 49 |
| Stage 2/100 years storage/ Stage 3 | – | – | – | – | – | – | – | 56 | 68 |

a) Original estimates from countries have been modified to correspond to 1 300 MW(e) unit size and to include contingency at 25 per cent.

The decommissioning costs for supporting nuclear fuel cycle facilities are reflected in the prices of these services, i.e., in the prices of enrichment, fabrication, reprocessing, etc. Hence, these costs transfer to the fuel cycle costs of nuclear power plants.

**The impact on electricity generating costs**

The impact of nuclear power plant decommissioning on the electricity generation costs can be indicated by calculating its share in the total levelised power production costs of the plant. Using the cost estimates provided by the countries, and a discount rate of 5 per cent (in real terms) it turns out that the funding of plant decommissioning generally accounts for less than 2 per cent of electricity generation costs. With lower discount rates the contribution is slightly larger and decommissioning can make up to 5 per cent of the total undiscounted costs related to constructing and operating a nuclear power plant over its lifetime. However, even allowing for uncertainties in the cost estimates and applicable discount rates, decommissioning contributes only a few per cent at most to electricity generation costs.

## Financing

Several countries have initiated schemes to assure that financing will be available for decommissioning activities, even if such activities are deferred for many decades after plants cease operation. Although it may be practical in some cases for utilities to pay decommissioning costs, when they occur, out of then current revenues, in most cases, countries have established some type of fund based on revenues paid in during the operating life of the plants. Details differ from country to country but generally such funds are based on having those who use the electricity pay all of its costs, present or future.

## Conclusion

Present decommissioning technology, previously demonstrated on small nuclear facilities, is deemed feasible for the larger commercial facilities that will require decommissioning in the future. There are incentives and ample scope for further reductions in worker exposure, waste volumes and costs. The development work ongoing in these areas adds to the already high level of confidence that nuclear facilities can be decommissioned safely and economically. Radioactive waste volumes resulting from the decommissioning of reactors and supporting fuel cycle facilities will likely be of the same order of magnitude as the amount of low and intermediate level waste generated by the operating reactors. The cost of decommissioning reactors and their supporting fuel cycle facilities are only a few per cent of the total electricity generating costs. Therefore, the decommissioning of commercial reactors and supporting fuel cycle facilities is considered to be technologically feasible, the waste volumes manageable and the costs affordable.

# 1. INTRODUCTION

Nuclear power plants and fuel cycle facilities are normally designed for an operating life-time of several decades. By appropriate refurbishment and upgrading of some equipment it is usually possible to continue the operations well beyond the design life-time. However, ultimately it becomes either technically or economically reasonable to shut down these facilities.

Decommissioning means all actions taken to retire the facility from service in a manner that provides for protection of the health and safety of the decommissioning workers, the general public, and the environment. These actions can range from merely closing down the facility with a minimal removal of radioactive material along with continuing maintenance and surveillance, to a complete removal of residual radioactivity to below a level determined acceptable for unrestricted use of the facility and site. This latter condition, unrestricted use, is the ultimate goal of all decommissioning actions at retired nuclear facilities unless the site is to be re-used for other nuclear purposes (1).

Several reports and studies published by various national and international organisations have discussed the economics and technical feasibility of decommissioning in general (1 and 2). The conclusion reached by those studies is that current technology can provide the decontamination, disassembly, radiation exposure control, and radioactive waste management needed to adequately decommission present-day nuclear facilities. This conclusion is based on past experience in decommissioning small test, training, and power demonstration reactors, nuclear support laboratories and non-reactor facilities, and on experience with relevant reactor repair and maintenance activities.

The present study focuses on the technical feasibility, needs, and costs of decommissioning the larger commercial facilities in the OECD member countries that are coming into service up to the year 2000. It is intended to inform the public and to assist in planning for the decommissioning of these facilities.

In this report, reactor decommissioning activities generally are considered to begin after operations have ceased and the fuel has been removed from the reactor, although in some countries the activities may be started while the fuel is still at the reactor site. Defuelling of gas-cooled reactors is a particularly lengthy process and may be done in parallel with initial decommissioning operations. However, the management of the spent fuel discharged at the end of operations is no different from the management of the fuel discharged during the reactor operating lifetime and its costs are a part of the total spent fuel management costs of that reactor. Therefore, this report does not deal with spent fuel management or its costs.

The three principal alternatives for decommissioning are described in Section 2 of this report. The factors to be considered in selecting the decommissioning strategy, i.e. a stage or a combination of stages that comprise the total decommissioning programme, are reviewed in Section 3. Section 4 contains a discussion of the feasibility of decommissioning techniques available for use on the larger reactors and fuel cycle facilities. The numbers and types of facilities to be decommissioned and the resultant waste volumes generated for disposal will then be projected in Section 5. Finally, the costs of decommissioning these facilities, the effect of these costs on electricity generating costs, and alternative methods of financing decommissioning are discussed in Section 6.

The discussion of decommissioning in this report draws on various countries' studies and experience in this area. Specific details about current activities and policies in NEA Member Countries are given in the short country specific Annexes at the end of the report.

The nuclear facilities that are addressed in this study include reactors, fuel fabrication facilities, reprocessing facilities, associated radioactive waste storage facilities, enrichment facilities and other directly related fuel cycle support facilities. The study does not include discussion on the decommissioning of uranium mines, mills, or mill tailings.

# 2. DESCRIPTION OF THE DECOMMISSIONING STAGES

The three principal stages of decommissioning have been previously defined in several reports (1, 3, and 4). The term *stage* as used herein, implies a particular set of conditions at the plant and does not necessarily imply that each stage must follow in sequence or that all stages are accomplished. In most cases the nuclear fuel and unattached radioactive materials in the process systems, as well as radioactive waste produced in normal operations, are removed prior to beginning the decommissioning. Each of the three decommissioning stages can be defined by two characteristics: the physical state of the plant and its equipment; and the surveillance necessitated by that state. Figure 1 illustrates the concept of these stages.

Although Stage 1 and Stage 2 can be considered as alternatives to each other, most countries consider them only as interim modes leading eventually to Stage 3.

## 2.1 STAGE 1 DECOMMISSIONING

**State of the plant and equipment**

For reactors, the first contamination barrier is kept as it was during operation but the mechanical opening systems are blocked and sealed (valves, plugs, etc.). The containment building is kept in a state appropriate to the remaining hazard. The atmosphere inside the containment building is subject to appropriate control of composition such as humidity and radioactivity. Access to the inside of the containment building is controlled, and personnel entering and exiting the containment are monitored for radioactivity. For fuel cycle facilities some of the mechanical operating systems may be retained for use during decontamination work in Stage 2.

**Surveillance**

The plant is kept under surveillance. The equipment necessary for monitoring radioactivity both inside the plant and in the area around the plant is kept in good condition and used when necessary and in accordance with national legal requirements. Routine inspections are carried out to assure that the plant remains in good condition. Checks are carried out to ensure that there are no leaks in the first contamination barrier or in the containment building.

# Figure 1. PRINCIPAL CHARACTERISTICS OF THE THREE DECOMMISSIONING STAGES FOR NUCLEAR POWER PLANTS

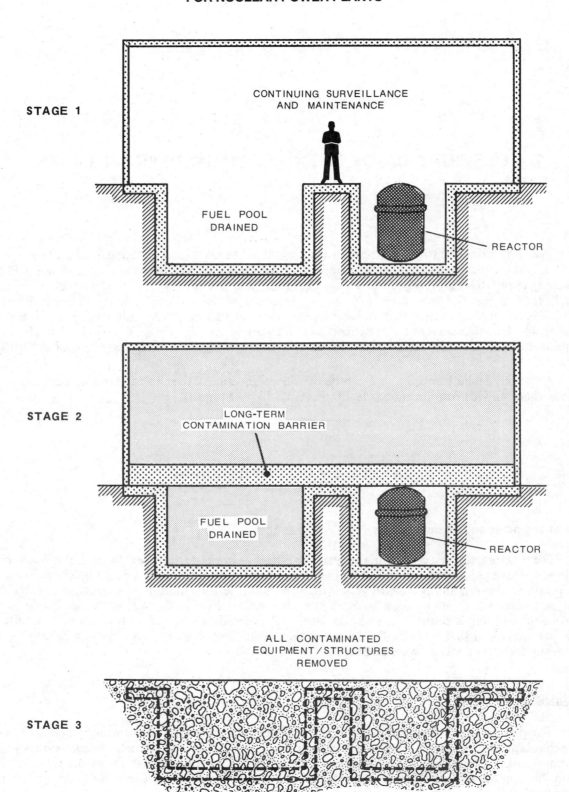

## 2.2 STAGE 2 DECOMMISSIONING

**State of the plant and equipment**

For reactors, the first contamination barrier is reduced to its minimum size (all parts easily dismantled are removed and the remaining barrier is sealed). The sealing of the barrier is reinforced by physical means and the biological shield is extended, if necessary, to completely surround the barrier. After decontamination to acceptable levels, the containment building and the ventilation system may be modified or removed if they no longer play a role in radiological safety. Depending upon the extent to which other equipment is removed or decontaminated, access to the containment building, if left standing, can be permitted. The non-radioactive parts of the plant (buildings or equipment) may be converted for new purposes. For fuel cycle facilities, the primary radioactive plant and equipment will sometimes be removed during Stage 2.

**Surveillance**

Surveillance around the barrier may be reduced but it is desirable for periodic spot checks to be continued, as well as surveillance of the environment. External inspection of the sealed parts is performed. Measurements for leaks are no longer necessary on the remaining containment building.

## 2.3 STAGE 3 DECOMMISSIONING

**State of the plant and equipment**

Materials, equipment and parts of the plant with activity levels significantly above natural background despite decontamination procedures, are removed. In all the remaining parts, contamination is below the authorised release limit, a very low level approaching that of the natural environment, unless the site is to be re-used for other nuclear purposes, e.g., for a replacement facility.

**Surveillance**

Unless re-used, the site is released without access restrictions due to residual radioactivity. From the point of view of radiological protection, no further surveillance is necessary.

# 3. FACTORS CONSIDERED IN THE SELECTION OF DECOMMISSIONING STRATEGY

The decision to retire a nuclear facility is generally made for economic reasons. In some cases, the refurbishment of the facility by replacing components which have reached the end of their operable lifetimes or the modifications which must be made to meet revisions in licensing regulations may be too costly in relationship to the expected revenues from the facility after modification. In other cases, a more efficient and less costly process may have been developed and the older facility cannot compete with the new process and cannot be modified to use the new process. In still other cases, the increased size and reduced unit costs from a larger facility may make a smaller existing facility uneconomical. A decision must then be made to decommission the facility in a manner which adequately continues the protection of the public health and safety, and the environment.

Factors which are important to consider in selecting the appropriate strategy of decommissioning for a particular nuclear facility include:

- National nuclear strategy
- Facility characteristics
- Protection of health, safety, and the environment
- Radioactive waste management
- Future use of the site
- Improvement of decommissioning technology
- Cost and availability of funds
- Social considerations.

The evaluations of these factors must obviously be weighed and balanced when selecting the decommissioning alternative (1 and 5). These evaluations should be made specifically for each nuclear facility. The weighing and balancing of the factors will vary from country to country and from case to case.

## 3.1 NATIONAL NUCLEAR STRATEGY

The selection of a decommissioning alternative must conform to the country's overall strategy and goals of nuclear power development. Questions of national strategy that are relevant to decommissioning include: the national policy for disposing of shutdown nuclear facilities (Is the disposal an industry or a government responsibility? Is there a plan for industry or government to carry out this responsibility?); the national policies for radioactive waste management and the national plan (strategy) for carrying out these policies (What is the plan and when will it be implemented? What is the policy for managing various types of waste?) and the national policies for worker and public health protection from radiation (What are the radiological dose limits for workers and the public? What are the policies for release of materials? What are the criteria to be used for decommissioning operations?). The national strategy will usually conform to international standards. The national nuclear strategy may change with time in some details. Some causes for a change in strategy might be technical developments, changes in regulations regarding radioactive waste disposal, changes in the availability of radioactive waste management facilities, changes in requirements for radiological protection for workers and the public, and changes to limits for release or disposal of material.

## 3.2   FACILITY CHARACTERISTICS

The characteristics, location, and total amount of the residual radioactivity in the facility are important elements in the selection of a decommissioning stage. Almost all the radioactivity in a reactor plant is contained in a small volume in the reactor vessel and internals that eventually makes 5-10 per cent of the radioactive waste arising from a Stage 3 dismantling (6). The radioactive waste from Stage 1 or Stage 2 decommissioning of a reactor plant is much less than from Stage 3, e.g. for a pressurised water reactor (PWR) it could be less than 10 per cent of the Stage 3 radioactive waste (2). For fuel cycle facilities the volume of waste from Stage 2 will be much less than from Stage 3 but if the primary radioactive plant and equipment are removed during Stage 2 the radioactivity will be much greater.

Radioactivity in a nuclear power plant arises principally from neutron irradiation of the reactor vessel and its internals along with a portion of the surrounding reinforced concrete (in a Canadian deuterium-uranium reactor (CANDU), which does not have a pressure vessel, the radioactivity is located principally in the pressure tubes and calandria). Some radioactivity is also produced by neutron irradiation of reactor control materials and the primary coolant piping adjacent to the reactor vessel. Contamination of the primary and auxiliary circuits arises from corrosion and erosion of these irradiated components and transport and deposition of the contaminants by the coolant. Contamination also arises due to neutron activation of corrosion products which are carried in the coolant through the reactor core. The coolant purification system becomes contaminated from the radioactive material which it removes from the reactor primary coolant system. Fuel element failures or leakage may also contribute to this contamination and leak fission products and transuranics into the reactor coolant system and in the fuel handling system. In pressurised light water reactors and CANDU heavy water reactors (HWR) the radioactivity is contained in the primary coolant system. In boiling water reactors (BWR) radioactivity is carried over in the steam to the turbine circuit resulting in active material depositing on the surface of the turbine condensers and feedwater system. In an advanced gas cooled reactor (AGR), the graphite moderator in the reactor contains significant quantities of carbon 14 radioisotope due to neutron irradiation. Similarly, in an HWR significant quantities of the radioisotope tritium are produced in the heavy water moderator.

The radioactive inventory is made up of both short-lived and long-lived radionuclides. If these radionuclides are principally short-lived isotopes in the range of 5 to 30-year half-lives (such as cobalt 60 and cesium 137), a significant reduction in radioactive inventory can be achieved by decommissioning the facility, initially, only to a safe storage condition to allow these radionuclides to decay before final dismantlement. This delay will tend to reduce occupational exposure to the decommissioning workers and may reduce the amount of material requiring disposal as radioactive waste. However, if there are significant quantities of the very long-lived alpha-emitting nuclides such as plutonium and other transuranics, as would be the case for a reprocessing plant or plutonium fuel-fabrication plant, there will be minimal advantage to delaying decommissioning from the standpoint of radiation protection.

Access for dismantlement of the reactor is restricted initially by the short half-life gamma emitter cobalt 60 for a time of about 60 to 90 years, depending upon the cobalt content of the steel from which the reactor was made (7 and 8). Beyond this time, the task is dominated by the radioactivity of nickel 63, silver 108, niobium 94 and nickel 59. Thus, deferral of dismantlement for periods up to about 90 years would facilitate the access and tend to reduce radiation doses to decommissioning workers if cobalt 60 is the dominant radionuclide. Beyond about 90 years little further benefit due to radioactivity reduction would be obtained.

Contamination of the peripheral buildings and equipment outside the biological shield is usually low enough to permit immediate access. If these buildings and equipment are decontaminated and removed soon after shutdown, the continuing effort needed for further maintenance and surveillance is greatly reduced. This is particularly significant if the building and equipment are in need of expensive repairs.

## 3.3 PROTECTION OF HEALTH, SAFETY AND THE ENVIRONMENT

A primary concern in any decommissioning program is to provide for the health and safety of the workers and to protect the general public and the environment. Regulations exist to assure that this protection is provided during nuclear facility operations and these regulations will also be applicable for decommissioning operations.

Experience has shown public exposures during decommissioning to be minimal (2), at most a few per cent of the exposure during reactor operation. Thus, public exposure is not likely to be a significant factor in selecting a decommissioning alternative.

All environmental impacts during the decommissioning of reactors and fuel cycle facilities are small and are within the regulation parameters for an operating facility (2). Therefore, they are not usually a governing factor in selecting a decommissioning alternative.

Protection of workers is an important consideration particularly in keeping with the principle of *As Low As Reasonably Achievable* (ALARA) for exposure reduction. Reduction of worker radiation dose can be achieved by deferring dismantlement to allow radioactive inventory to decay and by providing suitable remote-operated equipment, control procedures, and shielding. For fuel cycle facilities that handle plutonium or other long-lived radionuclides, the benefits from deferring dismantlement are, however, small. Worker dose reduction in dismantling these facilities may be achieved by decontamination of the facility, remote operations behind shielding, and by suitable control procedures.

## 3.4 RADIOACTIVE WASTE MANAGEMENT

At nuclear power plants the spent fuel must be removed from the reactor prior to completion of Stage 1 and especially before Stage 2 or Stage 3 can be started. Thus, the availability of facilities for reprocessing or storage of spent fuel away from the reactor may in some cases restrict the choice of decommissioning strategy, but often such facilities have to be made available earlier, during reactor operations.

The radioactive waste from reactor decommissioning is primarily low-level waste; a small amount of intermediate-level waste is also produced. The availability of off-site facilities for disposal of the low-level and intermediate-level decommissioning wastes is the most important factor in several OECD countries in making decisions on the timing of decommissioning a nuclear facility. The decommissioning waste can be disposed of in the same facilities as the low and intermediate level wastes that are continuously produced by the operating reactors. In fact (as will be seen in Section 5), in relation to these operations wastes, the amount of decommissioning wastes generated will be small over the next decade, although later their share of the total wastes will become more significant. For one reactor the volume of decommissioning wastes is estimated to be of the same order of magnitude as the volume of operations wastes produced during its lifetime.

Shallow land burial is permitted in some countries and these facilities are available to a limited extent. In other countries such as the Federal Republic of Germany (FRG) and Sweden, underground disposal is required, at least for intermediate-level waste. No such facilities are yet in operation, but these are being developed for use by the late 1980s. Ocean disposal has been used but has been suspended pending completion of further research and international agreement.

For decommissioning of plutonium fabrication facilities, the availability of facilities for storage, treatment and disposal of very long-lived alpha-bearing waste and intermediate-level waste from dismantling may affect the decision to select an alternative other than Stage 1 in some countries; similarly the availability of facilities for management of high-level waste and alpha wastes may affect the choice of the alternative for decommissioning a reprocessing facility. The decision may thus be linked to the development of storage and disposal facilities of spent fuel and high-level reprocessing wastes.

## 3.5   FUTURE USE OF THE SITE

Nuclear facility sites represent a valuable resource owing to their typically low seismic activity, proximity to cooling water supplies, access to an existing electrical distribution system, and local acceptance for nuclear power generation. Owners of nuclear facilities have come to recognise the value of locating replacement power or process facilities on the same location. If the site area is limited, the owner may have to dismantle the old nuclear power plant to make room for a replacement facility.

If the nuclear facility is adjacent to other facilities that will continue to remain in operation, Stage 1 decommissioning may be selected pending decommissioning of the other facilities. In this case, maintenance, surveillance, and security can be provided by personnel from the operating facility at little incremental cost.

## 3.6   FURTHER DEVELOPMENT OF DECOMMISSIONING TECHNOLOGY

Present technology or simple extensions thereto are adequate to accomplish any of the decommissioning alternatives safely. Nevertheless, development of this technology to reduce worker exposure, costs and waste volumes is continuing. Some delay in dismantlement may be advantageous to permit improved technology to be applied. For instance, the development of better equipment and methods for remote dismantlement is expected to reduce worker exposures. The development of better methods to rapidly discriminate between low levels of radioactivity and the development of equipment to consolidate wastes are expected to reduce the volume of radioactive waste. New equipment for dismantling facilities is expected to improve worker productivity and reduce costs.

## 3.7   COST AND AVAILABILITY OF FUNDS

The costs of the decommissioning option may be a very important factor. At the end of the plant operations, the immediate dismantling of the facility (Stage 3) would require a much larger expenditure than would be needed to accomplish Stage 1 only. If funds have been set aside for decommissioning during plant operation, then the immediate cost may be less important. The situation is different if the decommissioning has to compete for funds with the owner's other capital needs. Although the prediction of future decommissioning costs involves large uncertainties (see Section 6), plant owners should plan the financing of the chosen decommissioning alternative so that this item does not become a controlling factor.

## 3.8   SOCIAL AND OTHER CONSIDERATIONS

Various social considerations may affect the choice of decommissioning strategy. Examples of such issues are the effect on land values, aesthetic impact of the site, as well as changes in perceived risk associated with the continued existence of the shut-down plant on the site. Another important consideration, especially in areas where alternative employment is scarce, is that immediate dismantling requires a larger staff and work force than other alternatives and will result in a slower and smoother reduction of the operations staff.

# 4. TECHNOLOGY EXPERIENCE AND FUTURE DEVELOPMENT

The technology used to date in decommissioning smaller plants has proved to be quite satisfactory and demonstrated the feasibility of decommissioning to each of the three stages. Maintenance and repair work on the commercial operating reactors and post-accident decontamination and restoration have also provided experience. This technology is directly applicable to current and future decommissioning projects. Experience from these previous projects is important to guide the future development of decommissioning technology. Ongoing and planned decommissioning projects on progressively larger and more complex reactors and fuel cycle facilities will continue to add to this technical base. This Section reviews this experience, planned projects and areas for further technology developments.

## 4.1 DECOMMISSIONING EXPERIENCE WITH RESEARCH AND POWER DEMONSTRATION REACTORS AND FUEL CYCLE FACILITIES

Decommissioning experience to date has been mostly with research and power demonstration reactors which were of rather small size (less than about 300 MW(th) in power rating). All three stages of decommissioning involving the major types of reactors have been performed. Isolation of systems, handling of toxic as well as radioactive materials, use of controlled explosives for pipecutting and concrete demolition, use of various decontamination methods, remote segmentation of pressure vessel and internals, and surveillance practices comprise the major experience gained from these past decommissioning projects.

Some significant experience has been obtained in decommissioning of fuel cycle facilities. The Eurochemic reprocessing facility in Belgium was decontaminated and re-use of the facility for reprocessing is being considered. In the United States the Redox reprocessing facility was decontaminated to Stage 1 and the Westinghouse and General Electric Pu-U fuel fabrication facilities at Cheswick and Vallecitos, respectively, were decontaminated and glove boxes and equipment were sectioned and removed for Stage 3.

Selected decommissioning experiences from reactors and fuel cycle facilities are summarised in Table 1 to show their applicability to the commercial facilities to be decommissioned in the future.

## 4.2 REPAIR AND MAINTENANCE EXPERIENCE OF COMMERCIAL OPERATING NUCLEAR FACILITIES

Operating plants perform scheduled maintenance and necessary repair works every now and then. During these repair periods, a considerable emphasis is placed on keeping exposure as low as reasonably achievable (ALARA) while working under quite restrictive conditions in a cost-effective

Table 1

# SELECTED EXPERIENCE FROM DECOMMISSIONING RESEARCH
# AND POWER DEMONSTRATION REACTORS AND FUEL CYCLE FACILITIES
## (References 9 through 14)

### REACTORS[a]

| Facility name and location | Reactor type | Power rating | Decommissioning stage | Date | Experience |
|---|---|---|---|---|---|
| *United States* | | | | | |
| Carolina/Virginia Tube Reactor (CVTR) Parr, South Carolina | HWR | 65 MW(th) | Stage 1 | 1968 | Basic Stage 1 applicable to industrial scale reactor; procedure developed for periodic surveillance. |
| Pathfinder, Sioux Falls, South Dakota | BWR nuclear superheat | 190 MW(th) | Stage 1, conversion of facility to other use | 1972 | Isolation of steam plant and replacement of nuclear reactor with fossil fired boiler; continuous surveillance. |
| Saxton, Saxton, Pennsylvania | PWR | 23.5 MW(th) | Stage 1 | 1973 | Remote intrusion alarms for security to minimise work force. |
| Fermi I Monroe County, Michigan | Sodium cooled fast reactor | 200 MW(th) | Stage 1 | 1975 | Sodium handling experience for Stage 1. |
| Peach Bottom I York County, Pennsylvania | HTGR | 115 MW(th) | Stage 1 | 1978 | Core graphite fuel handling and disposal for Stage 1. |
| Hallam, Hallam, Nebraska | Graphite moderated, sodium cooled | 256 MW(th) | Stage 2 | 1968 | Basic Stage 2 procedures developed; no continuous surveillance. |
| Piqua Reactor, Piqua, Ohio | Organic cooled and moderated | 45 MW(th) | Stage 2 | 1969 | Entombment with conversion of reactor building to warehouse; reactor vessel entombed in sand; no continuous surveillance. |
| Boiling Nuclear Superheat Reactor (BONUS), Rincon, Puerto Rico | BWR with nuclear superheat | 50 MW(th) | Stage 2 | 1970 | Concrete entombment of vessel; decontamination of systems; release of site as exhibition center; no continuous surveillance. |
| Elk River Reactor (ERR) Elk River, Minnesota | BWR fossil-fuelled superheater | 58 MW(th) | Stage 3 | 1974 | Remote segmentation of vessel and internals; explosive demolition of concrete; survey and release of site for unrestricted use. |
| Sodium Reactor Experiment (SRE) Santa Susana, California | Graphite moderated, sodium cooled | 30 MW(th) | Stage 3 | 1983 | Remote segmentation of vessel and internals; explosive cutting of piping; release of site for unrestricted uses. |
| *France* | | | | | |
| Chinon A1 Reactor, Chinon | GCR | 300 MW(th) | Stage 1 | 1980 | Insulation of capacities from main piping to avoid moisture and radionuclide transfer. |
| EL 3 Saclay | HWR | 20 MW(th) | Stage 2 | 1986 | Development of tritiated waste embedment. |
| Pegase Reactor, Cadarache | LWR | 35 MW(th) | Stage 3 | 1978 | Decommissioned and transformed to a fuel storage facility. |
| *Federal Republic of Germany* | | | | | |
| Forschungsreaktor Neuherberg | | 1 MW(th) | Stage 2 | 1983 | Safe and irreversible enclosure of irradiated reactor components. |

## Table 1
### REACTORS *(cont'd)*

| Facility name and location | Reactor type | Power rating | Decommissioning stage | Date | Experience |
|---|---|---|---|---|---|
| Heissdampfreaktor Grosswelzheim | BWR | 100 MW(th) | Stage 1 | | Stage 1 decommissioning, remove segmentation of pressure-vessel internals, conversion of a reactor to a test facility. |
| NS Otto Hahn | PWR | 38 MW(th) | Stage 3 | 1981 | Stage 3 decommissioning of the NSSS of a ship, one-piece reactor vessel removal. |
| Niederaichbach Nuclear Power Station | GCHWR | 100 MW(e) | Stage 2 | 1981 | Stage 2 decommissioning and licensing. |
| *Sweden* | | | | | |
| Agesta | PHWR | 80 MW(e) | Stage 1 | 1975 | Procedure for Stage 1 decommissioning with limited surveillance. |
| *United Kingdom* | | | | | |
| BEPO Graphite-Moderated Reactor, AERE Harwell | GCR | 6.5 MW(th) | Stage 2 | | With emphasis on release of scrap metal for recovery. |

*a)*   See Glossary for explanation of abbreviations.

## Table 1 *(cont'd)*
### FUEL CYCLE FACILITIES

| Facility Name and Location | Facility type | Capacity | Decommissioning Stage | Date | Experience |
|---|---|---|---|---|---|
| *United States* | | | | | |
| Redox, Hanford | Reprocessing facility | Production size | Stage 1 | 1967 | Plutonium recovery programme conducted using various flushes. Systems drained and air dried. External flushing of equipment, cells and deck. Entrances locked. |
| Westinghouse, Cheswick, Pennsylvania | Pu-U fuel fabrication facility | Pilot-Size | Stage 3 | 1980-1983 | Decontamination, sectioning and removal of standard glove boxes, and equipment, storage tanks and the sintering furnace inside gloves boxes. |
| General Electric, Vallecitos, California | Pu-U fuel fabrication facility | Pilot-Size | Stage 3 | 1980-1982 | Similar to Westinghouse, Cheswick experience. |
| *Belgium* | | | | | |
| Eurochemic | Reprocessing facility | | Stage 2 and re-use | | Decontamination of process cells and equipment; decontamination, sectioning and removal of experimental gloveboxes and equipement. |
| *Federal Republic of Germany* | | | | | |
| Alkem | Pu-U fuel fabrication | Pilot-Size | Stage 3 | 1972 | Decontamination, sectioning and removal of glove boxes and equipment. |

manner. Effective techniques and procedures have been developed and proven in use to minimise plant down-time. Most of these proven, cost-effective techniques can be adapted and applied to similar activities during decommissioning.

### 4.2.1  Reactor Repairs

A wide variety of repair operations have been carried out at reactors around the world. These operations include: remote repair of small and large piping in LWRs and HWRs, fuel-channel pressure-tube replacement in HWRs, and tube repair (i.e. retubing in place) as well as replacement of steam generators in LWRs.

In Canada remote repair work has been done on a leaking heavy water pipe located in a vault underneath the Douglas Point reactor where radiation levels ranged up to almost 50 Gy/h. The work involved specially designed tooling, thorough training of repair crews, and full-scale mock-ups for testing, resulting in a low personnel radiation exposure.

Also, in Canada, calandria removal has been accomplished three times at the Chalk River large research reactors. After shut-down, removal of fuel and control mechanisms, draining and disconnection of systems, calandrias weighing from 3 500 kg to 11 560 kg were removed in one piece and transported to a nearby burial site for disposal. Extensive use of closed circuit TV, portable shields, and regular clean-up of contaminated areas kept personnel exposure to a minimum. Fuel-channel pressure-tubes have also been replaced in several Canadian reactors. The most significant of these activities has been the replacement of all pressure tubes in Pickering Units 1 and 2. This job, which has taken over two years, has been done primarily with simple manual tooling made from standard product forms and personnel radiation exposure was low. New automated techniques have been developed for use in any future retubing activities to further reduce personnel exposure and to reduce the time required.

Steam generator repair or replacement has been performed at several reactors. Steam generators have been replaced at the Point Beach, Robinson, Surry, and Turkey Point nuclear power plants in the United States and at Obrigheim in the Federal Republic of Germany. Work involved decontamination, disconnecting piping, lifting and the removal of the old steam generator, and replacement with a new one. All these operations were performed in relatively high radiation fields.

Oxidation problems in Magnox reactors in the United Kingdom led to the development of in-reactor inspection techniques and, subsequently, development of repair techniques, manipulators, remove control and visualisation. A range of repair techniques has been developed, based on the concept of self-contained packages which can be placed at the work location by existing manipulators to carry out the complete sequence of operations without the need to operate the manipulator.

In general, a seminar on Remote Handling Equipment held in Oxford in 1984 (15) concluded that remote handling equipment is very desirable for reducing the radiation exposure of personnel when applied to nuclear facilities during normal operation and repair. The main applications so far include visual inspection of rooms and components, radiation field measurements, weld inspection, vacuum cleaning, cleaning with aqueous solutions, by water lance, blasting, electro-polishing or flame-scarfing, cutting of metal tubes and plates by mechanical or thermal methods, and drilling of holes.

### 4.2.2  Fuel Cycle Facility Repairs

Publicly available, documented experience with non-reactor facility repair is somewhat limited, because there are fewer of these kinds of facilities and many of the facility processes are considered proprietary. With the exception of fuel reprocessing plants and plutonium fabrication facilities, non-reactor nuclear facilities have relatively low levels of penetrating radiation and more emphasis is placed on control of airborne contamination. Most operations can be performed *hands-on*, with little need for remotely operated equipment.

There has been experience gained in the decommissioning to Stages 1 or 3 of three small LWR fuel fabrication plants and seven facilities dedicated to plutonium fuel fabrication or related activities in

24

the United States. Several similar facilities have been decommissioned or are awaiting decommissioning in Europe and Asia.

Several fuel reprocessing plants have been shut down in the United States, Europe, and Asia. A few of these (Bombay, India; Dounreay, United Kingdom) have been decommissioned to a state approaching Stage 3. At another reprocessing plant (Belgium) main process installations have been decontaminated, and refurbishment of the plant is under consideration. In the United States the West Valley reprocessing plant is being decommissioned to Stage 3.

These various activities have provided information for the development of an information base that can be applied to decommissioning other similar industrial facilities in the future.

## 4.3   POST-ACCIDENT REFURBISHMENT OR DECOMMISSIONING OF REACTORS

Several reactors, after having sustained accidents, have been decontaminated and repaired for further operations, whereas others were decontaminated to some extent prior to being placed in a stage of decommissioning.

In the United States at the Enrico Fermi I fast breeder reactor, two fuel assemblies partially melted as a result of coolant flow blockage. Recovery began with the removal of all fuel assemblies, and draining of coolant for inspection of the pressure vessel. All activities were performed with specially designed remote equipment. With a new core in place, the reactor resumed operations briefly but was subsequently decommissioned to Stage 1 for economic reasons.

The Three Mile Island Nuclear Station Unit-2 in the United States experienced a loss of coolant accident which resulted in the uncovering and overheating of the reactor core. The fuel assemblies in the core underwent extensive damage, including melting of fuel, and volatile and gaseous radioactive fission products were released into the containment building through an open pressure relief valve. Decontamination has been accomplished in several heavily contaminated areas of the plant during the last several years to enable workers to proceed with other operations inside the containment building. Remote inspection of the reactor internals has been carried out using video cameras and long-reach tooling. Other remote equipment has been used for gathering samples in high radiation fields and is being constructed to remove the fuel fragments in the core. After fuel debris has been removed from the reactor vessel, the primary loop will be chemically decontaminated to remove remaining fuel pellet fragments and fission products which were distributed throughout the primary system during the accident. This project is providing very substantial knowledge which will be applicable to decommissioning.

Post-accident projects have shown that, even under conditions that are more severe than are found after normal operations, decontamination of equipment and areas, control of contamination, and worker exposure within the regulated dose limits can be successfully achieved. New equipment and procedures to deal with these specific post-accident problems has been very successful, though further development to reduce worker exposure and the cost of post-accident recovery is desirable.

## 4.4   ON-GOING AND PLANNED DECOMMISSIONING PROJECTS

The previous decommissioning repair, maintenance and post-accident experience is being used in planning and conducting current and forthcoming decommissioning projects. Many major decommissioning projects are either in progress, or are planned, as shown in Table 2. These projects are larger and more complicated than those previously undertaken. They will add substantially to the knowledge for Stage 1 and Stage 3 decommissioning and will provide a firmer basis for planning the decommissioning of the large commercial reactors and fuel cycle facilities.

## Table 2
## CURRENT AND PLANNED DECOMMISSIONING PROJECTS
### (Refs 10 and 16)

### REACTORS[a]

| Facility Name and Location | Reactor Type | Power Rating | Decommissioning Stage | Project Period | Expected Experience |
|---|---|---|---|---|---|
| **United States** | | | | | |
| Shippingport Atomic Power Station, Shippingport, Pennsylvania | PWR | 72 MW(e) | Stage 3 | 1985-1990 | Dismantling of reactor systems including one-piece reactor vessel removal. |
| Humboldt Bay Reactor, Humboldt Bay, California | BWR | 63 MW(e) | Stage 1 | 1983-1985 | Stage 1 for several decades. |
| Dresden 1 Reactor, Morris, Illinois | BWR | 207 MW(e) | Stage 1 | | Primary loop decontamination for larger BWR. |
| **France** | | | | | |
| Chinon A2, Chinon | GCR | 230 MW(e) | | | |
| EL4 Reactor, Brennillis | HWR | 70 MW(e) | | | |
| Rapsodie Reactor, Cadarache | LMFBR | 40 MW(th) | Stage 2 | 1985 | Dismantling an LMFBR. |
| G1 Reactor, Marcoule | GCR | 46 MW(th) | Stage 2 | 1989 | Decontamination to unrestricted release. |
| G2 Reactor, Marcoule | GCR | 40 MW(e) | Stage 2 | 1990 | Currently in Stage 2, will be proceeding to Stage 3 with removal and disposal of graphite internals. |
| G3 Reactor, Marcoule | GCR | 40 MW(e) | Stage 2 | 1994 | Currently in Stage 2, will be proceeding to Stage 3 with removal and disposal of graphite internals. |
| **Federal Republic of Germany** | | | | | |
| Niederaichbach Nuclear Power Station, Landshut, Bavaria | GCHWR | 100 MW(e) | Stage 3 | 1984-1990 | Stage 3 dismantlement of a reactor including cutting up of a reactor and structures. |
| Lingen Nuclear Power Station, Lingen | BWR | 240 MW(e) | Stage 1 | 1979 | Stage 1 decommissioning with surveillance and maintenance for about 30 years. |
| Gundremmingen Nuclear Power Station | BWR | 237 MW(e) | Stage 1 | 1983-1987 | Development of electrolytic decontamination with 90 % recovery of contaminated wastes, decontamination of most parts of the steam loop. |
| Mehrzwerckforschungs-reaktor, Karlsruhe | PHWR | 50 MW(e) | Stage 1 | 1984-1987 | Decommissioning of a heavy water cooled reactor. |
| FR2, Karlsruhe | HWR | 44 MW(th) | Stage 2 | 1982-1988 | Decommissioning of a heavy water cooled reactor, converting the reactor building into test laboratory. |
| **Italy** | | | | | |
| Garigliano Reator | BWR | 160 MW(e) | Stage 1 | | Stage 1 for 30 years. |
| **Japan** | | | | | |
| Japan Power Demonstration Reactor, (JPDR), Ibaraki | BWR | 12.5 MW(e) | Stage 3 | 1983-1990 | Development and demonstration of techniques for Stage 3. |

## Table 2
### REACTORS[a] *(cont'd)*

| Facility Name and Location | Reactor Type | Power Rating | Decommissioning Stage | Project Period | Expected Experience |
|---|---|---|---|---|---|
| *Canada* | | | | | |
| Gentilly-1 | PHWR | 250 MW(e) | Static State (between Stage 1 and Stage 2) | 1984-1986 | Decontamination using hydrolaser, dismantling of systems and components for building isolation, removal of asbestos, dry storage of spent fuel. |
| *United Kingdom* | | | | | |
| Windscale Reactor | AGR | 33 MW(e) | Stage 3 | 1981-1995 | Developing methods of thermal cutting (oxypropane plasma are cuttings). |
| *United States* | | | | | |
| West Valley, New York | Reprocessing facility for LWR fuel | 1 tonne/day | Stage 3 | 1982-1989 | Decontamination of facility, solidification of waste, glassification of waste. |
| Mound Lab, ANSPD Area, Ohio | Pu-238 Fabrication | | Stage 3 for re-use | 1978-1988 | Clean up and restoration of highly contaminated plutonium facility. |
| *France* | | | | | |
| AT1, La Hague | Reprocessing pilot plant for fast breeder fuel elements | 1 Kg/day | Stage 3 | 1982-1990 | Dismantling of highly contaminated plutonium glove boxes and hot cells. Waste handling procedures. |
| *United Kingdom* | | | | | |
| Windscale | Mixed Pu-U oxide production plant | 50 kg/day | Stage 2 | 1985-1989 | Decontamination and dismantling of highly contaminated plutonium equipment and glove boxes. |

a) See Glossary for explanation of abréviations.

## 4.5 APPLICATION OF EXPERIENCE TO COMMERCIAL REACTORS AND FUEL CYCLE FACILITIES

Present technology, although so far applied for decommissioning of small facilities only, is applicable for decommissioning of larger commercial-size facilities, also. The experience from decommissioning, decontamination and repair works will be used to plan for decommissioning these commercial reactors and fuel cycle facilities. Analysis of prior experience will permit projection of crew sizes and composition, productivity and material needs for such activities as pipe cutting, vessel segmentation and concrete demolition. Appropriate modifications will be made to account for the larger-size equipment involved in the commercial facilities, but the basic unit factors (dollars or manhours per cubic meter, manhours per pipe cut, etc.) developed from this experience will be useful for estimating future project costs, schedules, occupational exposures and waste volumes generated. Much work has been done and is continuing to extract these unit factors from the available historical data (17 and 18).

Modifications of the experience will also have to be made for commercial reactors because they have operated for longer periods and at higher power levels, which causes greater activation and higher radiation levels of metallic components after shutdown than experienced in smaller reactors. Some adjustment must be made in manhours projections to account for the lower labor productivity due to more remote operations and increased worker protection equipment. However, radiation from components is not expected to necessitate any major new or additional *difficulty factor* adjustments. The radiation levels of most components such as those in the primary loop are estimated generally to be low enough for direct personnel access after decontamination. The reactor vessel and internals, which have neutron-induced radioactivity, must be disassembled by remote operations in both small and large reactors.

## 4.6   TECHNOLOGY DEVELOPMENT

Prior studies and evaluations of decommissioning technology have concluded that decommissioning is certainly feasible with current technology (1, 2, 17 and 18). However, there are four primary areas where additional technology development is desirable to reduce radiation exposure, waste volumes, and costs, particularly for Stage 3 or for Stage 2 for fuel cycle facilities. These areas are:

1.   in-situ chemical decontamination of piping and components;
2.   disassembly operations including remote-operated equipment and tools for segmenting piping, components, and reactor vessel and internals;
3.   methods and equipment for waste treatment, volume reduction and efficient packaging; and
4.   improved measurement techniques to facilitate the segregation of different categories of waste and thereby reduce the volume of waste for disposal and hence also costs.

### 4.6.1   Decontamination

Chemical decontamination of reactor systems has been successfully demonstrated in decommissioning programmes as well as at operating commercial power plants for maintenance and refurbishment. In most cases, the objective has been to reduce radiation levels sufficiently to permit workers a longer access time for segmenting piping and components. Additional development and demonstration is desirable to further reduce radiation, contamination levels and waste volumes. In general, the processes now in use for maintenance of operating nuclear plants do not decontaminate surfaces to unrestricted access levels since care must be taken to avoid deterioration of a system that must be subsequently operated. However, decontamination for decommissioning does not need to be concerned with subsequent operation and therefore more vigorous reagents may be used. For instance, stronger acids can be employed which will remove more of the base metal as well as the surface contamination, thus achieving more effective decontamination.

Since most of the materials in the nuclear facilities are not activated but are contaminated on the surface, they can in principle be decontaminated to unrestricted release levels. However, the treatment needed must be proven cost-effective taking into account the time requirements and the amount of waste generated in decontamination. In addition, the management of the decontamination waste must not be adversely affected by the chemicals in the decontamination solutions (for example, chelating agents). Complete decontamination could make feasible the recovery or the unrestricted disposal of these materials and hence significantly decrease the amount of low level waste requiring disposal. It could also reduce consumption of some scarce metals. Methods for accomplishing complete decontamination while generating a minimum volume of waste are being developed such as electropolishing (a process which uses an electric current and a solvent to remove microscopic layers of the metal) and decontaminant gels (gel-based decontaminant compounds that are applied to the area to be decontaminated) (16 and 19). Decontamination could also be cost-effective by reducing the volume of high- and intermediate-level waste which is the most costly to dispose of.

28

### 4.6.2  Disassembly

The disassembly of commercial nuclear power plants will require segmentation of reactor vessels and internals to reduce these components to sizes that can be transported to disposal sites and to use the disposal space efficiently. Remote segmentation of the reactor vessel and internals was successfully demonstrated in decommissioning the Elk River Reactor using an oxyacetylene torch (20) and the Sodium Reactor Experiment (SRE) using a plasma arc torch (21). Additional development work is being accomplished in several countries to further automate the process using micro-computer control of cutter location, cutting parameters and verification of cut completion. Improvements are desirable for equipment setup and relocation as well as for removal of segmented sections of the vessel with reduced worker exposure. Work is underway in several NEA countries to develop mechanical cutting methods such as grinding or milling and thermal cutting methods.

Disassembly of piping and components has been successfully accomplished using generally-available industrial equipment (saw-blade, cutters, torches, shears, etc.) with appropriate modifications for contamination control (vacuum systems, enclosures, etc.). Research to develop remotely-operated equipment to combine crimping and arc sawing or shearing is underway in the United Kingdom to reduce the worker exposure for setup activities in high radiation zones.

Research to develop safer and more rapid techniques for removing asbestos insulation from pipe is being performed in the Federal Republic of Germany. An improved method employing a zipper-closed bag and pouch is being used for the Gentilly-1 decommissioning. The development of rapid rigging and handling fixtures to support pipe or components during cutting and subsequent transport of segments to a packaging/laydown area is desirable.

Demolition of heavily-reinforced concrete structures and scarification of contaminated concrete surfaces has been successfully accomplished using controlled blasting (Elk River Reactor), hydrauli-cally operated rams on a backhoe (Sodium Reactor Experiment), and drill and spall techniques (Hanford demonstration tests). However, in each case, the drilling device was a single-head, single drill (or ram) device using commercially available equipment. Further development is desirable to adapt multiple drill heads such as those used routinely in the mining industry with as many as eight heads per track drill. With the limited access available in power plants, it is not always possible to bring in more than one machine to drill. But a specially designed multiple head machine capable of withstanding the increased loads would reduce drilling time and operator exposure. Rebar cutting is generally done using an oxyacetylene torch for thick material, or using shears/bolt cutters for smaller gauge. Rapid, inexpensive cutters are being developed for thick rebar such as pre-formed explosive cutters or portable arc saws.

Much work has been done in the United Kingdom to develop large scale (2.5 meter) diamond saw cutting machines (22). Further work is desirable to develop methods for the quick removal of cut sections. Improvement in the process for cleaning out debris following blasting or ram-breaking might be made by adapting mining machinery for more rapid handling and removal of materials.

### 4.6.3  Waste Management

Low-level waste (LLW) from reactor operations and decommissioning has been safely disposed of by shallow land burial at controlled burial grounds, which are located in suitable retentive soils above the water table. This method has been used in the United States, France and the United Kingdom. Disposal on the ocean floor has been used but is suspended, pending further study by the London Dumping Conference.

The disposal of decommissioning waste can be facilitated by active development of volume reduction techniques. The decontamination of materials with surface contamination to a level suitable for unrestricted release serves the same purpose. Reduction of the reactor vessel and internals waste volumes can be accomplished by segmentation and techniques such as specially designed transport packaging for loading specific shapes into casks. This would offer maximum use of space in the casks and optimise the shielding. Loading of casks to mix higher activity segments with lower activity segments to maximise cask cavity utilisation is another technique now being studied in several countries such as the United States and the United Kingdom.

Volume reduction techniques for packaging piping and components include nesting of small bore piping in large diameter pipe where possible and slitting pipe longitudinally in thirds to lay sections flat in disposal boxes. This configuration quickly becomes mass-limited, necessitating the inclusion in these shipments of low-mass, high-volume materials such as insulation.

Other methods which are being employed worldwide to reduce the volume of LLW from operating reactors, can also be utilised to reduce the volume of LLW from decommissioning. One technique is the use of low-pressure (200-280 kPa) and high-pressure (500-550 kPa) compaction systems. These systems can achieve volume reduction ratios of 5:1 (low pressure) and 10:1 (high pressure) for appropriately segregated, non-metallic trash. Another technique is incineration of combustible radioactive materials. Incineration systems can achieve volume reduction ratios of as high as 250:1, although after solidification of scrub liquor sludge from the off gas system, the net reduction is closer to 25:1. Mobile incinerators are now beginning to enter the market for use at operating reactors and will be available for decommissioning. However, most of the material from decommissioning is neither compactable nor incinerable. Thus, the benefits from these methods are limited but may prove cost-effective in some instances.

A technique which is being developed in several countries is the melting and casting of metallic components into ingots. This technique achieves substantial volume reduction and also may contribute to the dilution of contamination and subsequently also retard the leaking of contaminants from the waste.

Some decommissioning waste types need further attention to develop waste management techniques and procedures. These waste types include tritiated wastes from heavy water reactors, mobile cesium wastes from decontamination, and high radioactivity waste from metallic reactor internals.

### 4.6.4   Exemption or Unrestricted Release Level

The exemption or unrestricted release level is that contamination level at which materials may be released for general use or disposal without further concern for residual radioactivity. A basic exemption criterion can be defined in terms of a potential dose rate to individuals who may use these materials or in terms of the risk of adverse health effects. The alpha, beta, and gamma radioactivity levels acceptable for unrestricted release of contaminated materials can be derived from this exemption criterion. These radioactivity levels are usually expressed as a specification for the residual radioactivity per unit of surface area (alpha, beta or gamma emissions per unit of time measured on a specified area). In the U.S. and some other countries, release of materials with residual surface contamination is done in conformity with the United States Nuclear Regulatory Commission Guide 1.86 (23). Because scrap materials are used in international commerce, there is need for international agreement on acceptable release levels of residual surface contamination. There is also a need for agreement on the level of radioactivity due to neutron activation that can remain in materials released for unrestricted use. This level would affect the volume of very slightly radioactive concrete rubble requiring disposal as radioactive waste rather than being used for on-site or off-site fill materials, and the re-use of materials recovered from reactors and fuel cycle facilities. In applying the exemption level, better technology is needed to rapidly measure radioactivity over large areas of materials.

## 4.7   CONCLUSIONS ON TECHNOLOGY EXPERIENCE AND FUTURE DEVELOPMENT

All three stages of decommissioning have been performed for small test, training, and power demonstration reactors and supporting fuel cycle facilities. Experience from decommissioning nuclear reactors already covers major reactor types. Some significant experience has been obtained in the decommissioning of fuel cycle facilities as well.

Present technology has proved to be quite satisfactory for decommissioning of nuclear facilities to any of the three stages. Though the experience to date is restricted to small facilities, the same technology is applicable to decommissioning of larger commercial-size facilities. Since the present-day reactors will have been operated for longer periods and at higher power levels than the reactors so far decommissioned, some adjustments in the working procedures are necessary to allow for the higher radiation levels, but the radiation from components is not expected to necessitate any essentially new approach.

Although current technology is sufficient for decommissioning of commercial-scale facilities, continued development in some areas is desirable. Such areas include decontamination methods, remotely operated equipment for facility and plant equipment disassembly, techniques for minimising waste generation through treatment, waste volume reduction, and discrimination of radioactivity levels in waste.

# 5. PROJECTIONS OF FACILITIES TO BE DECOMMISSIONED AND RESULTING WASTE VOLUMES

## 5.1 FACILITIES TO BE DECOMMISSIONED

To place the future decommissioning work in perspective, a cursory analysis was made on the number of nuclear power plants that could conceivably be decommissioned during the next forty years. This projection was subsequently used to estimate the volumes of radioactive waste that the nuclear facility decommissionings could give rise to.

The basis of the projections is the existing and projected nuclear power capacity in OECD countries in this century. In Appendix 1 the available data has been compiled to show the nuclear power capacity additions in successive 5-year periods through the year 2000. Although the actual reactor lifetimes will be variable, a crude estimate of the capacities retired in the future years can be obtained from the numbers shown by assuming a fixed lifetime for all reactors. Table 3 shows the results for a 25-year lifetime assumption. Although 25 years is commonly used as a reference lifetime, for instance, in economic calculations it is recognised that for most power plants the actual lifetimes are likely to become much longer and, indeed, may well be extended beyond 40 years by refurbishing and replacing components. Hence 25 years is a conservative assumption.

Table 3 shows that before the year 1995 the number of power plants that could conceivably require decommissioning is relatively small and their capacities are low in average, as may be seen from

Table 3

**NUMBER AND CAPACITY OF THE NUCLEAR POWER PLANTS REACHING 25-YEAR LIFETIME IN OECD COUNTRIES IN THE PERIOD 1981-2025**

| Reactor type | 1981-1995 | | 1996-2010 | | 2011-2025 |
|---|---|---|---|---|---|
| | No. of plants | Capacity GW(e) | No. of plants | Capacity GW(e) | Capacity GW(e) |
| PWR | 11 | 4.0 | 137 | 122.9 | |
| BWR | 8 | 3.2 | 67 | 57.2 | |
| HWR | 1 | 0.1 | 17 | 10.3 | |
| GCR | 31 | 4.9 | 5 | 2.2 | *a* |
| AGR | – | – | 9 | 5.1 | |
| HTR | – | – | 2 | 0.6 | |
| FBR | – | – | 3 | 1.6 | |
| Total | 51 | 12.2 | 240 | 199.9 | 140.8 |

*a)* Only total capacity estimated.

the estimate for the total capacity retired. The 25-year lifetime assumption implies that the number of power plants decommissioned would peak shortly after the turn of the century. In reality, such an effect is unlikely because plant operating lifetimes will most probably be extended and consequently a much slower increase in the number of decommissionings may be expected.

The non-reactor (i.e., fuel cycle) facilities are relatively few and the small volumes of decommissioning wastes generated from these facilities will not make a significant contribution to the total volume of waste requiring disposal from reactors. Quantitative projections were not attempted for the number of non-reactor facility decommissionings.

## 5.2  PROJECTED DECOMMISSIONING WASTE VOLUMES FOR DISPOSAL

Decommissioning of power reactors generates significant quantities of low-level wastes from neutron-activated materials and from the surface-contaminated materials. A small amount of intermediate-level waste comes from the reactor and its internals. Considering the total radioactivity of decommissioning wastes it should be noted, however, that more than 90 per cent of the radioactivity in a reactor at shutdown is contained in the spent fuel elements which will be removed and generally shipped off-site [the total activity of the spent fuel from a 1 000 MW(e) PWR shortly after shutdown is more than $10^7$ TBq (24); the activity of activated reactor components at shutdown is some $0.2 \times 10^6$ TBq, while other contributions to total activity inventory are small in comparison (17)].

Estimates of radioactive waste volumes arising from nuclear reactor decommissioning were received from Canada, the Federal Republic of Germany, Sweden and the United States (Table 4). All estimates are given for immediate dismantling of the reactor. The Canadian estimate is concerned with a four-unit PHWR representing a total of 2 060 MW of electric power capacity. The other estimates are for large PWRs and BWRs with sizes ranging from 800 MW(e) to 1 300 MW(e).

Table 4

**ESTIMATES ON THE VOLUMES OF LOW AND INTERMEDIATE LEVEL WASTE FROM REACTOR OPERTIONS AND DECOMMISSIONING ($m^3$)**

| Country | Canada | Federal Republic of Germany | | Sweden | | United States | |
|---|---|---|---|---|---|---|---|
| Size and type of reactor | 4 × 515 MWe PHWR | 1 200 MWe PWR | 800 MWe BWR | 900 MWe PWR | 1 000 MWe BWR | 1 000 MWe PWR | 1 000 MWe BWR |
| Wastes from 25-year operations[a] | 6 900-27 500 | 6 100-11 000 | 6 000-20 000 | 6 300 | 7 500 | 21 700 | 40 000 |
| Decommissioning wastes | 10 000 | 6 900 | 12 400 | 7 000 | 15 000 | 15 200 | 16 300 |
| Total wastes[a] (operations and decommissioning) | 16 900-37 500 | 13 000-17 900 | 18 400-32 400 | 13 300 | 22 500 | 36 900 | 56 300 |
| Decommissioning wastes as a fraction of total waste | 0.3-0.6 | 0.4-0.5 | 0.4-0.7 | 0.5 | 0.7 | 0.4 | 0.3 |

a)  Ranges in some estimates indicate the conceivable effect of possible incineration and compaction treatments.

The table shows differences in estimated waste amounts between reactor types and also between countries. The differences may reflect different plant designs, but the estimates are also affected by the assumed scope of volume reduction techniques and by the regulatory limits that determine which wastes are to be considered as radioactive wastes.

The information on waste volumes was used to derive an estimate for the total waste volumes from decommissioning in the next forty years. To obtain an upper estimate, immediate dismantling was assumed for all the reactors that, according to Table 3, were to reach 25-year lifetime during the three successive time periods considered.

The Canadian estimate, normalised per GW(e), was used as a basis for the projection of HWR decommissioning wastes. For LWRs, 15 000 m³ of radioactive waste was assumed to be produced by every GW(e) decommissioned. The results of the calculation are shown in Table 5. Before 1995 the decommissioning waste volumes are small, but could then, according to the 25-year lifetime assumption, increase significantly after the turn of the century. However, it must be remembered that even accepting the conservative lifetime assumption, the assumption of prompt dismantlement after shutdown for all reactors leads to an overestimate of waste volumes for the near-term.

The decommissioning wastes from GCRs, AGRs, HTRs and FBRs are not included in Table 5, due to lack of available data. The contribution from AGRs, HTRs and FBRs to total waste volumes would be negligible. The decommissioning of GCRs is not expected to increase the total waste volumes in 1981-1995, either, even though more than 60 per cent of the capacity retired during that period would, according to Table 3, be GCRs, because the United Kingdom currently plans to defer dismantling for 80 years or more to allow radioactivity to decay. Therefore, the inclusion of GCR waste volumes in the estimates would still not change the absolute waste volumes significantly.

Table 5

**RADIOACTIVE WASTE FROM DECOMMISSIONING LWR AND HWR POWER PLANTS IN OECD COUNTRIES IN THE PERIOD 1981-2025 ASSUMING PROMPT DISMANTLEMENT OF THE REACTORS AFTER 25 YEAR LIFETIME**

| Reactor type | Decommissioning waste produced per GW(e) | Total decommissioning waste produced Thousands of m³ | | |
|---|---|---|---|---|
| | Thousands of m³ | 1981-1995 | 1996-2010 | 2011-2025 |
| LWR | 15 | 110 | 2 650 | |
| HWR | 4.8 | 0.5 | 50 | |
| Total | | 110 | 2 700 | 2 200[a] |

a) Total volume estimated assuming 140 GW(e) to be decommissioned with 15 000 m³/GW(e) of waste generated.

As was mentioned earlier, the wastes from decommissioning can be disposed of in the same facilities as the low and intermediate level wastes that are continuously produced by the operating reactors. To provide a basis for comparison, estimates on the volumes of the operating wastes that need to be disposed of are also shown on Table 4. The estimates provided by the countries were normalised to show the volume of wastes from a 25-year operating period. The table shows that for all reactor types considered the volume of decommissioning waste is about the same order of magnitude as the volume of wastes from their operations during a 25-year period.

However, for a long time in the future most of the wastes actually generated and requiring disposal will originate from reactor operations. By assuming that some 240 m³/GW(e) of waste would be produced annually by LWRs and some 140 m³/GW(e) by HWRs (both numbers indicate the volumes after possible volume reduction treatments), some 900 000 m³ of low and intermediate level waste will have been generated by the reactor operations between 1981 and 1995. The amount of decommissioning waste produced by 1995 was estimated to be at most 100 000 m³, which is about 10 per cent of the total waste volume. Only later might the share of decommissioning wastes reach the values indicated in Table 4 on a reactor lifetime basis.

Although no projection was made on the number or type of fuel cycle facilities that will be decommissioned, an estimate was prepared by the U.S. of the typical waste volumes arising from a single

Table 6

**ESTIMATED VOLUMES OF RADIOACTIVE WASTE FROM COMMERCIAL NUCLEAR FUEL CYCLE FACILITIES** (24)

| Facility | Typical capacity | Reactor capacity supported | Volume and type of waste from decommissioning | | Volume of waste (LLW) from lifetime operations | Volume of decommissioning waste per GW(e) of reactors supported |
|---|---|---|---|---|---|---|
| | | GW(e) | m³ | type | m³ | m³ |
| Uranium conversion | 10 000 TML/year | 60 | 1 260 | LLW | 44 000 | 21 |
| Uranium enrichment | $8.75 \times 10^6$ SWU/year | 90 | 25 470 | LLW | 11 200 | 283 |
| Fuel fabrication | 1 000 THM/year | 33 | 1 100 | LLW | 99 000 | 33 |
| Reprocessing | 1 500 THM/year | 50 | 3 100 | LLW | 54 000 | 154 |
| | | | 4 600 | TRU | | |
| Total | | | | | | 491 |

THM  =  Tonne of Heavy Metal (uranium and plutonium).
LLW  =  Low Level Waste.
TRU  =  Transuranic Waste.

facility of each type. The information is shown in Table 6. In addition to the decommissioning waste volumes, it shows for comparison the amounts of wastes from their operating periods. To place the numbers in perspective, the decommissioning waste volumes from fuel cycle facilities can be compared with waste volumes from reactor decommissionings. Because a fuel cycle facility always supports several, perhaps a large number of reactors (typical reactor capacities supported are indicated in the table), a meaningful picture of the relative volumes is obtained by allocating the decommissioning wastes from a fuel cycle facility to all the reactors it supports. For example, from the decommissioning of a conversion plant some 21 m³ of waste is produced per 1 GW(e) of reactors supported. On a similar basis, all fuel cycle facilities considered in this report contribute together some 490 m³ of waste per 1 GW(e) of reactor capacity. This is less than 4 per cent of the U.S. estimate of 15 000 m³ for waste produced in the decommissioning of one PWR of about 1 GW(e) in size.

Some waste from decommissioning of fuel cycle facilities will be intermediate-level, high-level, or transuranic waste which will require special treatment. However, the storage and disposal facilities that are being developed to handle spent fuel and reprocessing waste can easily handle the small volumes of these decommissioning wastes as well, so that no specific facilities will be required.

Indeed, in decommissioning nuclear power plants, before proceeding further than Stage 1 spent fuel must be removed from the plant. In most cases, spent fuel will have been shipped off site during the reactor's operating lifetime for either reprocessing, long-term storage or disposal, according to national plans. In some countries long-term storage in water filled pools at the reactor site is utilised during the reactor lifetime. However, even in these cases, other long-term arrangements are being made and will be available by the time Stage 3 activities are ready to start. In the meantime, Stage 1 activities can be carried out where needed.

Facilities will also be provided for low level and intermediate level waste from decommissioning of reactors and fuel cycle facilities. Facilities for intermediate level waste are being developed. Planning and construction of facilities with adequate capacity to handle the low-level waste volumes from decommissioning is in progress. The technology is available to handle the waste from both operations and decommissioning. In those countries where there is a problem of low-level waste disposal, the problem is one of licensing and, since it will have to be resolved to handle operating wastes in any case, a satisfactory solution should be available by the time significant volumes of decommissioning wastes are produced.

## 5.3  CONCLUSIONS ON WASTE ARISINGS

The wastes from reactor decommissioning are primarily low-level wastes and can be disposed of in the same or similar facilities as the wastes from reactor operations. The volume of radioactive decommissioning wastes from one reactor is of the same order of magnitude as the volume of wastes from its lifetime operations. Wide experience already exists on the handling of reactor wastes and disposal facilities for such wastes are being developed in many countries. The disposal of decommissioning wastes from reactors will not require new technical approaches.

The high-level or transuranic decommissioning waste from fuel cycle facilities will require special attention. However, the facilities being developed for spent fuel and reprocessing wastes can accommodate the small volumes of these decommissioning wastes as well. In general, fuel cycle facilities are relatively few and the small decommissioning waste volumes they generate will not make a significant contribution to the total waste volumes.

# 6.  COSTS AND FINANCING OF DECOMMISSIONING

## 6.1  COSTS OF DECOMMISSIONING

Decommissioning costs are a function of three interrelated factors as follows:
- the sequence of decommissioning stages chosen,
- the timing to accomplish each decommissioning stage,
- the decommissioning activities accomplished in each stage.

In addition, cost comparisons between various decommissioning strategies are affected by the discount rates used in calculations. All cost estimates for future decommissioning activities include considerable uncertainties. These uncertainties will gradually diminish with growing experience from current and planned projects.

This discussion of decommissioning costs deals principally with nuclear power plants, since it will be shown that the decommissioning costs for the nuclear fuel cycle facilities are a small fraction of the decommissioning costs of the nuclear power plants supported by them. In any case, the costs for decommissioning nuclear fuel cycle facilities are normally included in the price of fuel and are therefore paid from the power plant electricity generating revenues during the plant lifetime.

### 6.1.1  Cost Factors

Total decommissioning costs in this Section mean all costs that arise during the time period from the start of decommissioning until the moment when the site is released for unrestricted use. It is assumed that neither Stage 1 nor Stage 2 completes decommissioning but each will be followed eventually by Stage 3, because the radioactivity from certain long-lived radionuclides will remain above the limits for unrestricted access to the site for a time period much longer than the durability of the Stage 2 containment (25). Hence, the total decommissioning cost will be the cost of Stage 3 immediately after shutdown or the sum of the costs of either Stage 1 plus a storage period plus a delayed Stage 3, or Stage 2 plus a storage period plus a delayed Stage 3.

Cost estimates are based on previous experience with decommissioning small nuclear facilities, with maintenance and component replacements in large nuclear power plants and with similar non-nuclear work. Costs depend on many country – and site-specific factors such as:
- type of nuclear power plant;
- waste transportation and disposal practices;
- labor rates.

The sequence and timing of various decommissioning stages may have important implications on the total decommissioning costs. For reactors, the deferment of Stage 3 will reduce the amount of decontamination, remote operations, and worker protection required. Labor productivity should be increased and the amount of radioactive waste for transport and disposal may be reduced, even though the latter effect may be small as long as the radioactivity in materials has not decayed below the levels for unrestricted release. These effects result in lower costs during Stage 3. Countering the reduced cost for dismantlement resulting from the reduced radioactivity is the cumulative annual storage costs following Stage 1 or Stage 2. During storage there are fixed annual costs for maintenance, security, insurance, and possibly licensing fees, which, over long periods, may exceed the savings from lower radiation levels. The optimum balance has to be developed on a case- by-case basis because of the differences between facilities due to their type, design and operating history, which affect the types and amounts of radioactivity and, consequently, the labor productivity, shielding and remote operations needed, and amount of waste. Another factor which affects labor productivity is the availability of skilled plant operating personnel for decommissioning; the personnel who operated the plant are not likely to be available if Stage 2 or Stage 3 are delayed for many years. Some factors such as labor rates and waste disposal costs, are country-specific, as are also the anticipated inflation and interest rates, which are reflected in the discount rate used in the comparative calculations. In a wider cost-benefit perspective, the value of possible worker dose reduction may be taken into account in the search for optimum strategy.

It is assumed that plans will have been made to provide adequate funds for decommissioning so that funding should not generally be a constraint on choosing a decommissioning strategy. However, for a few early plants, this may not be true.

Storage periods of up to one hundred years are being contemplated for nuclear power plants. For the cost estimates in this report, a storage period of 30 years was chosen after Stage 1 as a minimum period to obtain significant decay of radioactivity, and 100 years was chosen after Stage 2, as a period long enough to allow worker access generally without the need for shielding or remote operations. However, for fuel cycle facilities which handle only long-lived radionuclides such as uranium and plutonium, radioactive decay is so slow that there is no benefit to delay from the viewpoint of worker dose reduction.

### 6.1.2 Cost Estimates For Nuclear Power Plant Decommissioning

Estimates for power reactor decommissioning costs were obtained from the U.S., the Federal Republic of Germany, Sweden, Finland and Canada. Estimates were given for various decommissioning strategies leading to unrestricted access of the site. For the purposes of comparisons the original cost estimates were scaled by the ratio of the power capacities to correspond to a 1 300 MW(e) unit. If not originally included, all estimates were adjusted to contain 25 per cent for contingencies.

The modified estimates are summarised in Table 7. All figures shown represent undiscounted total decommissioning costs in constant January 1984 U.S. dollars for the strategies indicated. A list of exchange rates applicable as of 1st January, 1984, is included as Appendix 4. The original estimates as submitted by the countries can be found in Appendix 2.

The original cost estimates from the United States corresponded to a 1 175 MW(e) pressurised water reactor and a 1 155 MW(e) boiling water reactor. The estimates prepared by the Battelle Pacific Northwest Laboratory for the United States Nuclear Regulatory Commission (17 and 18) were updated for this study and are similar to those given in Reference 26. The estimates for Stage 3 include dismantlement and removal of all structures, including uncontaminated buildings. The estimates do not include the costs which may be incurred during the period from cessation of power production and the start of decommissioning when some final plant operations, such as shipping spent fuel offsite are conducted. Some other estimates that have been made of decommissioning costs are summarised in Reference 27 and also in the U.S. country Annex at the end of this report; a cost of $130 million in 1984 U.S. dollars is considered to be an average with a range of $80 to $200 million.

The estimates from the Federal Republic of Germany are based on the costs for decommissioning the 1 200 MW(e) PWR Biblis A plant and the 800 MW(e) BWR Brunsbüttel plant (28). The original contingency allowances were 20 per cent for PWRs and 35 per cent for BWRs; for this report these were changed to 25 per cent for uniformity of assumptions.

The Swedish estimates for immediate Stage 3 decommissioning were originally for BWRs ranging in power capacities from 440 MW(e) to 1 050 MW(e), and for a 915 MW(e) PWR (29). The original estimates included a 25 per cent contingency factor.

The Finnish estimates were for the Loviisa Nuclear Power Station with two 445 MW(e) PWRs, and the Olkiluoto Nuclear Power Station with two 710 MW(e) BWRs. For the PWRs, the decommissioning strategy considered was immediate dismantling to Stage 3; for the BWRs the strategy was Stage 1 followed by 30 years storage and Stage 3. However, dismantling is assumed to be restricted to contaminated parts only (the sites will remain in power plant use). The original contingency allowances in the estimates were 30 per cent for PWRs and 20 per cent for BWRs.

The estimates for the Canadian pressurised heavy water reactors are based on estimated costs for Pickering A power station with 4 units of 515 MW(e) each (30 and 31). Considered are both immediate dismantling and dismantling after 30 years storage. The original contingency allowance was 10 per cent.

There is a relatively good agreement between the estimates for PWR, while for BWR the range of estimates is larger. However, it must be noticed that the linear normalisation of costs to correspond to the same plant size is likely to distort the picture, and the distortion may be significant when the difference to the reference size is large or when costs for multi-unit plant are compared with costs for a single unit. The modification of contingency allowances may lead to errors, since the differences in the original contingency factors may reflect a different degree of conservatism or a different degree of precision that has already been built into the cost factors. The underlying definitions of decommissioning stages may also differ from country to country. The estimates in Table 7 cannot be considered as national estimates for a 1 300 MW(e) facility. However, they give a sense of the range of actual decommissioning costs for large power reactors.

Table 7

**TOTAL UNDISCOUNTED COSTS FOR VARIOUS DECOMMISSIONING STRATEGIES**
ORIGINAL ESTIMATES FROM COUNTRIES HAVE BEEN MODIFIED TO CORRESPOND
TO 1 300 MW(e) UNIT SIZE AND TO INCLUDE CONTINGENCY AT 25 %

Millions of January 1984 U.S. dollars

|  | Canada | Federal Republic of Germany | | Finland | | Sweden | | United States | |
| --- | --- | --- | --- | --- | --- | --- | --- | --- | --- |
|  | HWR | PWR | BWR | PWR | BWR | PWR | BWR | PWR | BWR |
| Stage 3 immediately | 145 | 119 | 173 | 105 | – | 107 | 140 | 97 | 113 |
| Stage 1/30 years storage/Stage 3 | 117 | 121 | 181 | – | 126 | – | – | 121 | 141 |
| Stage 2/100 years storage/Stage 3 | – | – | – | – | – | – | – | 158 | 186 |

In both United States and Federal Republic of Germany the cost estimate for immediate Stage 3 decommissioning is lower than the estimate for Stage 1 followed by Stage 3 after 30 years. However, the comparison on the basis of Table 7 is valid only if no emphasis is laid on the fact that in the case of immediate dismantling the costs occur during a few years after reactor shutdown, whereas in the other cases the costs are spread over a long time period. Table 8 shows the comparison in the case where all costs are discounted by an annual rate of 5 per cent to the year of reactor shutdown. In other words, the

Table 8

**TOTAL DISCOUNTED COSTS FOR VARIOUS DECOMMISSIONING STRATEGIES**
ORIGINAL ESTIMATES FROM COUNTRIES HAVE BEEN MODIFIED
TO CORRESPOND TO 1 300 MW(e) UNIT SIZE
AND TO INCLUDE CONTINGENCY AT 25 %
(5 % DISCOUNT RATE; COSTS DISCOUNTED TO THE YEAR OF REACTOR SHUTDOWN)
Millions of January 1984 U.S. dollars

| | Canada | Federal Republic of Germany | | Finland | | Sweden | | United States | |
|---|---|---|---|---|---|---|---|---|---|
| | HWR | PWR | BWR | PWR | BWR | PWR | BWR | PWR | BWR |
| Stage 3 immediately | 129 | 105 | 153 | 93 | – | 95 | 124 | 86 | 100 |
| Stage 1/30 years storage/Stage 3 | 29 | 30 | 44 | – | 29 | – | – | 41 | 49 |
| Stage 2/100 years storage/Stage 3 | – | – | – | – | – | – | – | 56 | 68 |

*Figure 2.* **THE IMPACT OF DISCOUNTING ON THE ESTIMATED TOTAL DECOMMISSIONING COSTS**
(According to the U.S. cost data)

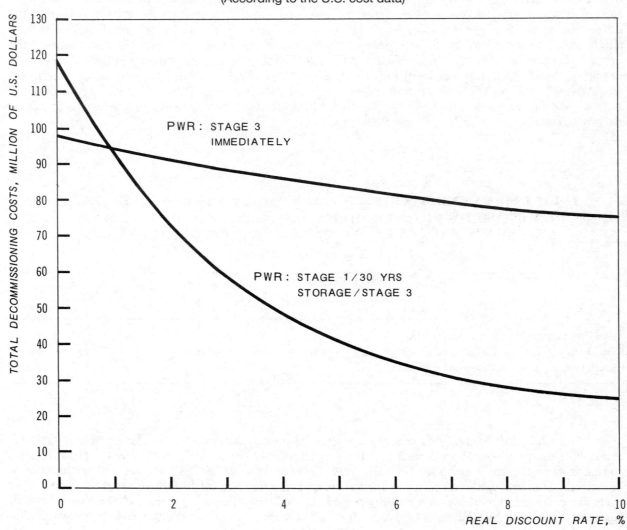

numbers of Table 8 represent the amount of funds that would be required for decommissioning in the year of reactor shutdown if a 5 per cent real interest rate (interest above inflation) could be subsequently realised for the funds until they are needed. It is seen that, allowing for discounting, in all cases delaying Stage 3 would be economically preferable to undertaking it immediately. Figure 2 shows that for the U.S. cost estimates this conclusion is in fact valid for all discount rates above about 1 per cent. [For background and details of discounting methodology the reader is invited to study the text in Appendix 3.]

### 6.1.3  Nuclear Fuel Cycle Facilities Decommissioning Costs

Fuel cycle facilities for nuclear power plants may include:
–  uranium mines and mills,
–  uranium refineries and conversion plants,
–  uranium enrichment plants,
–  fuel fabrication plants,
–  spent fuel reprocessing plants, and
–  heavy water manufacturing plants (only for heavy water reactors).

The mining and milling facilities are outside the scope of this study. The costs to decommission these facilities are, as a rule, inluded in the prices of their products and, therefore, cannot be included in the nuclear power plant decommissioning costs. Their decommissioning costs were, nevertheless, calculated as a matter of interest, based primarily on the studies by Pacific Northwest Laboratories (32 through 35). The decommissioning cost of an enrichment plant was derived by analogy with comparable activities in decommissioning a reactor plant. When prorated over the number of reactors which each nuclear fuel cycle facility serves (see Table 6), the costs for Stage 3 decommissioning of the supporting fuel cycle facilities in the United States when allocated to a 1 000 MWe LWR were, in 1984 U.S. dollars:

$230 000/reactor for $UF_6$ plant
$160 000/reactor for U enrichment plant
$140 000/reactor for fuel fabrication plant
$2 800 000/reactor for reprocessing plant

TOTAL  $3 330 000/reactor for all fuel cycle facilities.

Thus, the per-reactor share of the cost of decommissioning fuel cycle facilities is small compared to the cost of decommissioning the reactors they support.

## 6.2  THE IMPACT OF DECOMMISSIONING ON ELECTRICITY GENERATION COSTS

### 6.2.1  Calculation of the Cost Impact

A useful measure of unit electricity generation costs is obtained by asking which constant sum of money should be charged for each kilowatthour of electricity produced through the lifetime of the facility, to precisely cover all the costs that the production of electricity by this facility entails. These costs include the costs of building and operating the plants as well as the costs of waste management and facility decommissioning. Altogether they are spread over a long period of time, possibly more than one hundred years. To obtain a meaningful measure that can be easily compared with other costs, all costs must be expressed in constant money of a given reference year, i.e., assuming for the monetary unit the same purchasing power (that of the given year) throughout the period considered. However, since money can normally be invested to produce revenue, the calculation of the charge must be adjusted to allow for differences in the times at which the costs are paid and the charges collected. This is done by requiring that the sum of discounted revenues for electricity must equal the sum of discounted costs for its production when a constant sum of money is charged for each unit of electricity produced and when the same reference year and the same discount rate are used for discounting both costs and revenues.

The measure corresponding to these requirements is called the levelised cost of electricity generation. Its magnitude depends on the discount rate chosen for the calculation, but it should be noted that to be consistent with the constant money basis, the discount rate must correspond to the real rate of interest, i.e., the rate of interest above inflation. According to the information gathered for the study of Reference 36, discount rates of about 5 per cent are considered in most OECD countries appropriate for generation cost calculations, although somewhat higher rates are used in a few countries.

The levelisation procedure may also be used to determine the costs of decommissioning per kilowatthour of electricity produced, when it is required that each kilowatthour produced during the reactor lifetime must bear an equal fraction of the costs. The levelised decommissioning cost represents the amount of money that, if charged for each kilowatthour produced, would be sufficient for the complete recovery of the decommissioning costs at the time these occur assuming that until that time the charges collected return interest at the rate specified for discounting. The effect of decommissioning on generating costs is briefly discussed in the following chapter. More details of the calculations and the methodology used may be found in Appendix 3.

### 6.2.2 Levelised Decommissioning Costs

Estimates for decommissioning costs as calculated per kilowatthour of electricity produced are shown in Table 9. The estimates have been calculated from the data discussed in Section 6.1 assuming 25-year reactor lifetime, 73 per cent average load factor and 5 per cent annual discount rate. The three decommissioning strategies are the same as in Tables 7 and 8; their timing is explained in detail in Appendix 3.

Table 9
#### DECOMMISSIONING COSTS PER UNIT OF ELECTRICITY PRODUCED
(25-YEAR UNIT LIFETIME, 5 POUR CENT DISCOUNT RATE)

(January 1984 mills/khw)

|  |  | Stage 3 immediately | Stage 1/30 years Stage 3 | Stage 2/100 years Stage 3 |
|---|---|---|---|---|
| United States | PWR | 0.2 | 0.1 | 0.1 |
|  | BWR | 0.2 | 0.1 | 0.2 |
| Federal Republic of Germany | PWR | 0.3 | 0.1 |  |
|  | BWR | 0.4 | 0.1 |  |
| Sweden | PWR | 0.2 |  |  |
|  | BWR | 0.4 |  |  |
| Finland | PWR | 0.2 |  |  |
|  | BWR |  | 0.1 |  |
| Canada | HWR | 0.3 |  |  |

In all cases considered the estimated impact on power generation costs is found to be small, at most some 0.4 mills/kWh and only some 0.1 mills/kWh for all cases where Stage 1 and 30 years delay precede Stage 3. Consistent with Table 8, allowing for the different times the expenditures occur (through discounting) one generally obtains smaller costs for delayed decommissioning strategies than for immediate dismantling.

In perspective, the levelised decommissioning costs for all the strategies considered represent a very small fraction of the total power generation costs. The comparison in Table 10 is based on a parallel study of power generation costs in OECD countries (36). It shows the fraction which the decommissioning costs per kilowatthour are of the total power generating costs in the country considered. In all countries studied, the decommissioning is seen to represent at most about 1.5 per cent of the total power generation costs. The variation in the percentage figures is considerably larger than in the mills/kWh figures of Table 9. The reason for this is the differences in generating costs: for the countries included in Table 10 the levelised generating costs varied between 20.3 and 43.8 mills/kWh according to the study that was used as a reference (36).

Table 10

**LEVELISED DECOMMISSIONING COSTS AS A FRACTION OF THE TOTAL
LEVELISED POWER GENERATION COSTS, PER CENT**

5 PER CENT DISCOUNT RATE, 25-YEAR LIFETIME

In percentage

|  |  | Stage 3 immediately | Stage 1/30 years Stage 3 | Stage 2/100 years Stage 3 |
|---|---|---|---|---|
| United States | PWR[a] | 0.5 | 0.3 | 0.2 |
| Federal Republic of Germany | PWR | 1.1 | 0.4 | |
| Sweden | PWR[b] | 0.4 | | |
| Finland | PWR | 0.8 | | |
| Canada | HWR[c] | 1.5 | 0.5 | |

a) Plant located in the central U.S.
b) Generation costs data from the 1983 generation cost report of the NEA.
c) Plant located in central Canada.

A shorter reactor lifetime would imply a higher cost per kilowatt-hour for decommissioning, assuming that the total costs of decommissioning would still be the same as for the 25-year lifetime. For immediate Stage 3 dismantling and assuming 5 per cent discount rate, the cost per kilowatt-hour for 20-year lifetime might be up to 50 per cent higher than in the case of 25-year lifetime. Increasing the reactor lifetime from 25 years would have an opposite effect. In both cases the effect on the relative figures of Table 10 would be smaller, since shorter lifetime would cause higher total generation costs while longer lifetime would reduce these.

The choice of discount rate may have a large effect on results. Increasing the discount rate from the 5 per cent used in Tables 9 and 10 would result in lower levelised decommissioning costs, and the share of decommissioning in the total levelised power generating costs would also be smaller, since higher discount rates in general lead to higher levelised power generating costs for nuclear power plants. For low discount rates the decommissioning costs per kilowatthour would increase, and the increase is largest for the strategies involving delays before Stage 3. In absolute terms, without discounting the decommissioning cost per kilowatt-hour would approach 1 mill for all strategies and reactor types considered.

Since the study of Reference 36 does not include results for levelised power generation costs at zero discount rate, any direct comparison with undiscounted decommissioning costs is impossible. In fact, since capital investments normally imply interest charges or in the case of equity financing, return on investment, the calculation of levelised generation costs with zero discount rate may not be relevant. In Table 11 the undiscounted decommissioning costs are compared with the total undiscounted costs related to the building and operating of the nuclear power plant. Interest during construction is included

Table 11

**UNDISCOUNTED DECOMMISSIONING COSTS AS A FRACTION
OF THE TOTAL UNDISCOUNTED LIFETIME COSTS OF NUCLEAR POWER PLANTS
INTEREST DURING CONSTRUCTION INCLUDED IN TOTAL PLANT COSTS
AT 5 % PER ANNUM**

In percentage

|  |  | Stage 3 immediately | Stage 1/30 years Stage 3 | Stage 2/100 years Stage 3 |
|---|---|---|---|---|
| United States | PWR[a] | 1.6 | 2.0 | 2.6 |
| Federal Republic of Germany | PWR | 2.7 | 2.8 | |
| Sweden | PWR[b] | 1.7 | | |
| Finland | PWR | 2.7 | | |
| Canada | HWR[c] | 4.7 | 3.8 | |

a) Plant located in the central U.S..
b) Generation costs data from the 1983 generation cost report of the NEA.
c) Plant located in central Canada.

at 5 per cent per year in real terms. The figures represent rough estimates based on data of Reference 36. It is seen that even in this case the decommissioning expenditures correspond to only a few per cent of the total costs that the production of electricity at that plant gives rise to.

## 6.3   FINANCING OF DECOMMISSIONING

### 6.3.1   Planning

The financing of decommissioning of nuclear reactors is complicated by the potentially long delay periods between plant shutdown and final decommissioning. Depending on the decommissioning option chosen, this time may be up to about 100 years.

Methods of assuring that financing will be available for decommissioning are still evolving and different countries have taken different approaches. If the utility is assured of surviving as a financially viable institution in the long run, for example if it is government owned, it may be feasible for it to pay for decommissioning at the time the costs are incurred out of its then current revenues. However most countries appear to have opted for some method of raising funds during the plant's operating life, sufficient to cover expected eventual decommissioning costs. This method seeks to assure that those using the electricity from the plant pay for all of its costs, present and future.

The exact future cost of decommissioning is difficult to estimate but the estimates can be updated, and the rate of funding adjusted over the plant life, as new information becomes available. The estimates used in establishing the fund are unlikely to be exact, but the fund could be expected to cover at least the major part of decommissioning. This would minimise the cost burden to future generations and could, if estimates prove to be too high, even leave them with a windfall.

Decommissioning funds need to be invested in a manner which offers a high degree of security that they will be available when required. The real interest rate assumed in establishing annual decommissioning charges should be consistent with realistic expectations of interest paid on such secure investments.

The operating lifetime of the plant has to be estimated when a funding scheme is established. Generally, conservative (short) lifetime assumptions are used for this purpose. If plants are run for longer times the charge for the decommissioning fund could be reduced or even eliminated in later years of operation. In some cases some form of insurance may be taken out to cover the risk of underfunding due to a plant shutting down prematurely owing to unforeseen circumstances such as a serious accident.

The system used for funding decommissioning will vary from country to country depending on circumstances. No one system is likely to be right for all countries, and some evolution in means of assuring availability of funding is likely to occur over the next several years at least.

## 6.3.2  Examples of Approaches

In most OECD countries financial plans exist or are under consideration for decommissioning of nuclear power plants. In some countries the facility owners may each decide on the practical approach, while in some other countries a common policy and detailed financing rules have been established.

In Canada facility owners may decide on the best financing method. The approach currently applied is that when the costs of removal of fixed assets can be reasonably estimated, the amounts are charged to operations over the remaining service life of the asset.

In France decommissioning costs are treated similarly to the construction costs.

In Finland the utilities are obliged to set aside funds for decommissioning. The amount of annual provisions made is proportional to the electricity production. Funds are presently kept with the electricity producers, but the proposed new atomic energy legislation would require establishment of an external fund.

In the Federal Republic of Germany, funds are collected during the operational phase of the nuclear power plant within a period of time which represents the useful lifetime of the steam generating system (37). Decommissioning costs are estimated based on the studies of the Biblis A and Brunsbüttel plants (28). Details of funding have to be decided upon by the financial authority for each plant individually.

In Sweden, the nuclear power plant owner pays an annual fee which is related to the energy produced. The fee is determined annually by the government based on updated cost calculations provided by the reactor owner. The collected fees are deposited in interest-bearing accounts, one for each reactor owner, with the National Bank of Sweden. The power plant owner may borrow from the collected fees. When the power plant is being decommissioned and dismantled the owner will be reimbursed from the collected fees for these costs. This same procedure is used for other activities for the *back-end* of the fuel cycle.

In the United States, the Nuclear Regulatory Commission has proposed that a decommissioning fund be set aside for each nuclear facility which will be adequate to terminate the facility license by decommissioning to Stage 3. For a large nuclear power plant the fund would be about $100 million. Access to the fund by the facility owner would be controlled and limitations would be placed on the investments that could be made with the fund (38). For nuclear power plants, the Public Utility Commissions in each state will decide how the funds may be collected in the rate base.

## 6.4  CONCLUSION ON COSTS AND FINANCING

It has been shown in this section that decommissioning costs have a very small impact on the electricity generation costs of a large nuclear power plant. In levelised power production costs its share is a few per cent at most. The uncertainties in estimating the decommissioning costs, applicable real discounting rates or facility operating lifetime are large but still very unlikely to significantly affect that conclusion. The conclusion is based on calculations made for light water reactors and pressurised heavy water reactors but it is likely to be applicable to other types of reactors as well. Several methods of financing these future decommissioning costs can be used with varying degrees of cost effectiveness, assurance of funding availability when needed, and equity. The choice of a method for financing will be highly dependent upon the circumstances of each utility and the country in which it operates.

# REFERENCES

1. *Decommissioning of Nuclear Facilities: Decontamination, Disassembly and Waste Management*, Technical Reports Series No. 230, International Atomic Energy Agency, Vienna, Austria, 1983.

2. *Draft Generic Environmental Impact Statement on Decommissioning of Nuclear Facilities*, U.S. Nuclear Regulatory Commission, January 1981 (NUREG-0586).

3. *Decommissioning of Nuclear Facilities*, Technical Document, IAEA-179, International Atomic Energy Agency, Vienna, Austria, 1975.

4. *Factors Relevant to the Decommissioning of Land-Based Nuclear Reactor Plants*, Safety Series No. 52, International Atomic Energy Agency, Vienna, Austria, 1980.

5. *Storage with Surveillance vs. Immediate Decommissioning for Nuclear Reactors*. Proceedings of an NEA Workshop, OECD Nuclear Energy Agency, Paris, 1985.

6. *Technology, Safety and Costs of Decommissioning a Reference Pressurised Water Reactor Power Station*, Battelle Pacific Northwest Laboratories, September 1984 (NUREG/CR-0130 Add. 3).

7. Boothby, R.M., and Williams, T.M., *The Control of Cobalt Content in Reactor Grade Steels*, Nucl. Sci. Technol., Vol. 5 (1983), No. 2 (EUR 8655 EN).

8. Goddard, A.J.H., et. al., *Trace Element Assessment of Low-Alloy and Stainless Steels with Reference to Gamma Activities*, Commission of the European Communities, Brussels, 1984, (EUR 9264 EN).

9. Erikson, Peter, D., *U.S Licensed Reactor Decommissioning Experience*, U.S. Nuclear Regulatory Commission, American Nuclear Society Topical Meeting, September 16-20, 1979.

10. *Compendium on Decommissioning Activities in NEA Member Countries*, OECD Nuclear Energy Agency, Paris, January 1985.

11. *Directory of Nuclear Reactors*, Vol. VI, International Atomic Energy Agency, Vienna, 1966, p. 221.

12. *Directory of Nuclear Reactors*, Vol. II, International Atomic Energy Agency, Vienna, 1960, p. 257.

13. *Directory of Nuclear Reactors*, Vol. II, International Atomic Energy Agency, Vienna, 1959, p. 295.

14. *International Data Collection and Analysis*, Vol. II, Nuclear Assurance Corporation, June 1978.

15. *Proceedings of the Seminar on Remote Handling Equipment for Nuclear Fuel Cycle Facilities*, Oxford, 2-5 October 1984, OECD Nuclear Energy Agency, Paris, 1985.

16. *Decommissioning of Nuclear Power Plants*, Proceedings of a European Conference held in Luxembourg, 22-24 May 1984. Commission of the European Communities.

17. *Technology, Safety and Costs of Decommissioning a Reference Pressurised Water Reactor Power Station*, Pacific Northwest Laboratory for U.S. Nuclear Regulatory Commission, 1978 (NUREG/CR-0130).

18. *Technology, Safety and Costs of Decommissioning a Reference Boiling Water Reactor Power Station*, Pacific Northwest Laboratory for U.S. Nuclear Regulatory Commission, 1980 (NUREG/CR-0672).

19. *Decontamination of Nuclear Facilities to Permit Operation, Inspection, Maintenance, Modification or Plant Decommissioning*, Technical Report Series No. 249, International Atomic Energy Agency, 1985.

20. *Final Program Report – AEC Elk River Reactor*, United Power Association, September 1974, (COO-651-93).

21. *Sodium Reactor Experiment Decommissioning Final Report*, Rockwell International Energy Systems Group, August 15, 1983 (ESG-DOE-13403).

22. Rawlings, A.W., *Development of Large Diamond-tipped Saws and their Application to Cutting Large Radioactive Reinforced Concrete Structures*, Commission of the European Communities, Brussels, 1985 (EUR 9499 EN).

23. *Regulatory Guide 1.86, Termination of Operating Licenses for Nuclear Reactors*, U.S. Atomic Energy Commission, June 1974.

24. *Spent Fuel and Radioactive Waste Inventories, Projections and Characteristics*, U.S. Department of Energy, 1984 (DOE/RW-0006).

25. Evans, J.C. et al, *Long-Lived Activation Products in Reactor Materials*, Pacific Northwest Laboratory, August 1984 (NUREG/CR-3474).

26. *Updated Costs for Decommissioning Nuclear Power Facilities*. Electric Power Research Institute, May 1985 (EPRI NP-4012).

27. Nuclear Energy Cost Data Base, U.S. Department of Energy, June 1985 (DOE/NE-0043/3).

28. Watzel G.V.P., et al, *Decommissioning of Nuclear Power Stations in the Federal Republic of Germany at the End of Their Service Life*, Progress Report of the Association of German Engineers, Series: Environmental Technology, Series 15, No. 18, 1982.

29. *Technology and Costs for Dismantling a Swedish Nuclear Power Plant*, Swedish Nuclear Fuel and Waste Management Co., 1979, (SKBF/KBS TR 79-22).

30. *Preliminary Nuclear Decommissioning Cost Study*, Ontario Hydro Report No. 81156, April 1981.

31. *Decommissioning by Immediate Dismantlement, Preliminary Cost Estimate for Pickering Nuclear Generating Station A*, Ontario Hydro, Report No. 82208, July 1982.

32. Jenkins, C.E. et al, *Technology Safety, and Costs of Decommissioning a Reference Small Mixed-Oxide Fuel Fabrication Plant*, Pacific Northwest Laboratory, February 1979 (NUREG/CR-0129).

33. Elder, H.K., Blahnik, D.E., *Technology, Safety and Costs of Decommissioning a Reference Uranium Fuel Fabrication Plant*, Pacific Northwest Laboratory, October 1980 (NUREG/CR-1266).

34. Elder, H.K., *Technology, Safety and Costs of Decommissioning a Reference Uranium Hexafluoride Conversion Plant*, Pacific Northwest Laboratory, June 1981 (NUREG/CR-1757).

35. Schneider, K.G. et al, *Technology, Safety and Costs of Decommissioning a Reference Nuclear Fuel Reprocessing Plant*, Pacific Northwest Laboratory, October 1977 (NUREG-0278).

36. *Projected Costs of Generating Electricity from Nuclear and Coal-Fired Power Stations for Commissioning in 1995*, OECD Nuclear Energy Agency, 1986.

37. Reinhard, H., *Die Bildung von Rückstellungen für die Kosten der Stillegung und Beseitigung von Kernkraftwerken*, Energiewirtschaftliche Tagesfragen 32 (1982), Heft 8.

38. U.S. Government Federal Register, Vol. 50, No. 28, pg. 5600, February 11, 1985. *Decommissioning Criteria for Nuclear Facilities*.

# GLOSSARY

**Absorbed Dose Rate:** The increment of absorbed dose in a particular medium during a given time interval.

**Activity:** The activity, A, of an amount of radioactive nuclide in a particular energy state at a given time is the quotient of dN by dt, where dN is the expectation value of the number of spontaneous nuclear transformations from that energy state in the time interval dt:

$$A = \frac{dN}{dt}$$

The special name for the SI unit of activity is becquerel (Bq):
1 Bq = 1 s$^{-1}$. The special unit of activity, curie (Ci), may be used temporarily: 1 Ci = 3.7 × 10$^{10}$ Bq (exactly).

**Advanced Gas-Cooled Reactor (AGR):** A gas-cooled reactor using slightly enriched uranium fuel.

**ALARA:** *As low as reasonably achievable, economic and social factors being taken into account.* A basic principle of radiation protection taken from the Recommendations of the International Commission on Radiological Protection (ICRP), ICRP Publication No. 26.

**Alpha-Bearing Waste:** Waste containing one or more alpha-emitting radionuclides, usually actinides, in quantities above acceptable limits. The limits are established by the national regulatory body.

**Barrier** (natural or engineered): A feature which delays or prevents radionuclide migration from the waste and/or repository into its surroundings. An engineered barrier is a feature made by or altered by man; it may be part of the waste package and/or part of the repository.

**Biological Shield:** Physical barriers to reduce exposure to living organisms, e.g., the concrete shield around the reactor.

**Boiling-Water Reactor (BWR):** A light-water reactor in which water, used as both coolant and moderator, is allowed to boil in the core. The resulting steam can be used directly to drive a turbine.

**Breeder Reactor:** A reactor that produces more fissionable fuel than it consumes. The new fissionable material is created by a process known as breeding, in which neutrons from fission are captured in fertile materials.

**CANDU:** Canadian Deuterium-Uranium reactor. A pressurised heavy water reactor of Canadian design, which uses natural uranium as a fuel and heavy water as a moderator and coolant.

**Capacity (Electrical):** The load for which a generating unit is rated, either by the user or by the manufacturer.

**Cask** (or flask): A transport container providing shielding for radioactive materials and holding one or more canisters.

**Containment:**   The retention of radioactive material in such a way that it is effectively prevented from becoming dispersed into the environment or only released at an acceptable rate.

**Contamination, Radioactive:**   A radioactive substance in a material or place where it is undesirable. Surface contamination is the result of the deposition and attachment of radioactive materials to a surface.

**Decommissioning:**   The actions taken at the end of a facility's useful life for its planned permanent retirement from active service. (See also Stage of Decommissioning.)

**Decontamination:**   Removal or reduction of radioactive contamination physically by removing the surface itself or chemically by removing surface films containing radioactive materials.

**Dismantlement or Disassembly:**   Those actions required to disassemble and/or remove radioactive or contaminated materials from the facility and site.

**Disposal:**   The emplacement of waste materials in a repository, or at a given location, without the intention of retrieval. Disposal also covers direct discharge of both gaseous and liquid effluents into the environment.

**Dose:**   A term denoting the quantity of radiation energy absorbed by a medium divided by the mass of the medium.

**Exposure:**   Any exposure of persons to ionising radiation. A distinction is made between:

    *a)*   external exposure, being exposure to sources outside the body;
    *b)*   internal exposure, being exposure to sources inside the body;
    *c)*   total exposure, being the sum of the external and internal exposures.

**Facility:**   The physical complex of buildings and equipment within a site. Fuel cycle facilities are facilities used in the preparation of fuel materials for use in nuclear power reactors.

**Fast Breeder Reactor (FBR):**   A reactor in which the fission chain reaction is sustained primarily by fast neutrons rather than by thermal or intermediate neutrons. Fast reactors require little or no moderator to slow down the neutrons from the speeds at which they are ejected from fissioning nuclei. Also, this type of reactor produces more fissile material than it consumes.

**Gas-Cooled Heavy Water Reactor (GCHWR):**   See Heavy Water Reactor.

**Gas-Cooled Reactor (GCR):**   A graphite-moderated, carbon-dioxide-cooled reactor.

**Gigawatt (GW):**   One million kilowatts. GW(e) refers to electric power, whereas GW(th) refers to thermal power.

**Half-Life, Radioactive:**   For a single radioactive decay process, the time required for the activity to decrease to half its value by that process. (After a period equal to ten half-lives, the activity has decreased to about 0.1 per cent of its original value).

**Heavy-Water Reactor (HWR):**   A reactor using heavy water as moderator.

**High-Level Waste:**   The highly radioactive waste material that results from the reprocessing of spent nuclear fuel, including liquid waste produced directly in reprocessing and any solid waste derived from the liquid and which contains a combination of TRU waste and fission products in such concentration as to require long-term isolation.

**High-Temperature Reactor (HTR or HTGR):**   A graphite-moderated, helium-cooled advanced reactor.

**Interim Storage (Storage):**   A storage operation for which:

    *a)*   monitoring and human control are provided, and
    *b)*   subsequent action involving treatment, transportation, and final disposition is expected.

**Intermediate-Level Waste (or medium-level waste):**   Waste of a lower activity level and heat output than high-level waste, but which still requires shielding during handling and transportation. The term is used generally to refer to all wastes not defined as either high-level or low-level. (See alpha-bearing waste and long-lived waste for other possible limitations.)

**Light-Water Reactor (LWR):**   A nuclear reactor that uses ordinary water as the primary coolant and moderator, with slightly enriched uranium as fuel. There are two types of commercial light-water reactor: the boiling-water reactor (BWR) and the pressurised-water reactor (PWR).

**Long-Lived Nuclide:**   For waste management purposes, a radioactive isotope with a half-life greater than about 30 years.

**Low-Level Waste:**   Waste which, because of its radionuclide content, does not require shielding during normal handling and transportation. (See alpha-bearing waste and high-level waste for other possible limitations.)

**Megawatt (MW):**   One thousand kilowatts (see also Gigawatt).

**Monitoring:**   Taking measurements or observations for recognising the status, or significant changes in conditions or performance, of a facility or area.

**Nuclear Fuel:**   Fissionable and/or fertile material for use as fuel in a nuclear reactor. Fissionable materials can be fissioned by neutrons. Fertile material can be converted to fissionable material by absorbing neutrons.

**Nuclear Power Plant:**   A single- or multi-unit facility in which heat produced in a reactor(s) by the fissioning of nuclear fuel is used to drive a steam turbine(s) which in turn drives an electric generator.

**Nuclear Reactor:**   An apparatus in which the nuclear fission chain can be initiated, maintained, and controlled so that energy is released at a specific rate. The reactor apparatus includes fissionable material (fuel) such as uranium or plutonium; fertile material; moderating material (unless it is a fast reactor); a containment vessel; shielding to protect personnel; provision for heat removal; and control elements and instrumentation.

**Pressurized Heavy-Water Reactor (PHWR):**   See Heavy Water Reactor.

**Pressurized-Water Reactor (PWR):**   A nuclear reactor in which heat is transferred from the core to a heat exchanger via water kept under high pressure, so that high temperatures can be maintained in the primary system without boiling the water. Steam is generated in a secondary circuit.

**Physical Decontamination or Mechanical Decontamination:**   Decontamination accomplished by the use of mechanical cleaning means or by the removal of the surface itself.

**Plant:**   The physical complex or buildings and equipment, including the site.

**Radioactive Material:**   A material of which one or more constituents exhibit radioactivity.

**Radioactive Waste:**   Any material that contains or is contaminated with radionuclides at concentrations or radioactivity levels greater than the *exempt quantities* established by the authorities and for which no use is foreseen.

**Radioactivity:**   The property of certain nuclides of spontaneously emitting particles or gamma radiation, or of emitting X-radiation following orbital electron capture, or of undergoing spontaneous fission.

**Reactor Internals:**   Internal reactor components such as the fuel support structure, the coolant flow distribution components and neutron reflector or absorber (shielding) components.

**Regulatory Authority or Regulatory Body:**   An authority or system of authorities designated by the Government of a Member State as having the legal authority for conducting the licensing process, for issuing of licenses and thereby for regulating the siting, design, construction, commissioning, operation, shutdown, decommissioning and subsequent control of nuclear facilities (e.g. waste repositories) or specific aspects thereof. This authority could be a body (existing or to be established) in the field of nuclear-related health and safety or mining safety or environmental protection, vested with such legal authority, or it could be the Government or a department of the Government.

**Safe Storage:**   A period of time starting after the initial decommissioning activities of preparation for safe storage cease and in which surveillance and maintenance of the facility takes place. The duration of time can vary from a few years up to an order of 100 years. (See stage of decommissioning).

**Shallow-Land Disposal** (e.g. shallow-land burial): Disposal of radioactive waste, with or without engineered barriers, above or below the ground surface, where the final protective covering is of the order of up to about 10 meters thick. Some Member countries consider *shallow-land disposal* to be a mode of storage rather than a mode of disposal.

**Stage of Decommissioning:** The term *stage*, implies a state or condition of a facility after decommissioning activities:

Stage 1 – storage with surveillance;
Stage 2 – restricted site release;
Stage 3 – unrestricted site release.

**Surveillance:** Includes all planned activities performed to ensure that the conditions at a nuclear installation remain within the prescribed limits; it should detect in a timely manner any unsafe condition and the degradation of structures, systems and components which could at a later time result in an unsafe condition. These activities may comprise:

a) monitoring of individual parameters and system status;
b) checks and calibrations of instrumentation;
c) testing and inspection of structures, systems and components;
d) evaluation of the results of items a) and c).

**Testing:** Determination or verification of the capability of a component or assembly of components to meet specified requirements by subjecting the component or assembly to a set of physical, chemical, environmental or operational conditions.

**Transuranic (TRU) Waste:** Waste containing quantities of nuclides above agreed limits having atomic numbers above 92. The limits are established by national regulatory bodies.

**Treatment of Waste:** Operations intended to benefit safety or economy by changing the characteristics of the waste. Three basic treatment concepts are:

a) volume reduction;
b) removal of radionuclides from the waste;
c) change of composition.

**Waste Arisings:** Radioactive wastes generated by any stage in the nuclear fuel cycle.

**Waste Disposal:** See disposal.

**Waste Management:** All activities, administrative and operational, that are involved in the handling, treatment, conditioning, transportation, storage and disposal of waste.

# APPENDICES

# NUCLEAR POWER CAPACITY ADDITIONS 1960-2000

## NUMBER OF NEW PLANTS – CAPACITY ADDITIONS (GWe)

| Country | Reactor Type | Before 1960 | 1961-1965 | 1966-1970 | 1971-1975 | 1976-1980 | 1981-1985 | 1986-1990 | 1991-1995 | 1996-2000 |
|---|---|---|---|---|---|---|---|---|---|---|
| Belgium | PWR | – | – | – | 3- 1.7 | 1- 0.9 | 3- 2.9 | – | – | – |
| Canada | HWR | – | – | – | 4- 2.1 | 4- 3.1 | 8- 4.5 | 8- 5.7 | – | – |
| Finland | PWR | – | – | – | – | 1- 0.5 | 1- 0.5 | – | 1- 1.0 | – |
|  | BWR | – | – | – | – | 1- 0.7 | 1- 0.7 | – | – | – |
| France | PWR | – | – | 1-0.3 | – | 14-12.7 | 25-24.7 | 12-14.6 | } a-10.3 | } a-10.8 |
|  | GCR | 2-0.1 | 1-0.2 | 2-0.3 | 2- 1.0 | – | – | – |  |  |
|  | FBR | – | – | – | 1- 0.2 | – | 1- 1.2 | – |  |  |
| Federal Republic of Germany | PWR | – | – | 1-0.3 | 2- 1.8 | 3- 3.3 | 3- 3.7 | 5- 6.3 | – | } a- 1.4 |
|  | BWR | – | – | – | 1- 0.6 | 3- 2.5 | 3- 3.8 | – | – |  |
|  | FBR | – | – | – | – | – | – | 1- 0.3 | – |  |
|  | HTR | – | – | – | – | – | 1- 0.3 | – | – |  |
| Italy | PWR | – | 1-0.3 | – | – | – | – | – | 2- 2.0 | 4- 4.0 |
|  | BWR | – | – | – | – | 1- 0.8 | – | 2- 2.0 | – | – |
|  | GCR | – | 1-0.2 | – | – | – | – | – | – | – |
| Japan | PWR | – | – | 1-0.3 | 4- 2.6 | 4- 3.5 | 5- 3.6 | 3- 2.6 | } a-15.3 | } a-13.4 |
|  | BWR | – | – | 1-0.3 | 3- 1.7 | 7- 5.7 | 5- 4.8 | 5- 5.0 |  |  |
|  | GCR | – | – | 1-0.2 | – | – | – | – |  |  |
|  | HWR | – | – | – | – | 1- 0.2 | – | – |  |  |
|  | FBR | – | – | – | – | – | – | – | 1- 0.3 | – |
| Netherlands | PWR | – | – | – | 1- 0.45 | – | – | – | } a- 1.0 | } a- 1.0 |
|  | BWR | – | – | 1-0.05 | – | – | – | – |  |  |
| Spain | PWR | – | – | 1-0.2 | – | – | 4- 3.6 | 4- 1.9 | } a- 1.8 | } a- 1.2 |
|  | BWR | – | – | – | 1- 0.4 | 1- 0.9 | – | – |  |  |
|  | GCR | – | – | – | 1- 0.4 | – | – | – | – | – |
| Sweden | PWR | – | – | – | 1- 0.8 | – | 2- 1.8 | – | – | – |
|  | BWR | – | – | – | 3- 1.6 | 3- 2.2 | 3- 3.0 | – | – | – |
| Switzerland | PWR | – | – | 1-0.4 | 1- 0.3 | 1- 0.2 | – | – | – | – |
|  | BWR | – | – | – | 1- 0.3 | – | 1- 1.0 | – | 1- 0.9 | – |
| Turkey | PWR | – | – | – | – | – | – | – | 1- 1.0 | } a- 1.2 |
|  | HWR | – | – | – | – | – | – | – | 1- 0.6 |  |
| United Kingdom | PWR | – | – | – | – | – | – | – | – | } a- 5.4 |
|  | HWR | – | – | 1-0.1 | – | – | – | – | – |  |
|  | GCR | 8-0.3 | 12-2.1 | 4-0.9 | 2- 0.8 | – | – | – | – |  |
|  | AGR | – | – | – | – | 4- 2.1 | 5- 3.0 | 5- 3.1 | – |  |
|  | FBR | – | – | – | – | 1- 0.2 | – | – | – |  |
| United States | PWR | – | 1-0.2 | 4-2.0 | 23-17.6 | 15-13.8 | 20-21.9 | 16-19.2 | 2- 1.2 | 2- 2.2 |
|  | BWR | – | 1-0.1 | 5-2.7 | 15-12.1 | 3- 2.7 | 11-11.7 | 3- 4.1 | – | – |
|  | HTR | – | – | – | – | 1- 0.3 | – | – | – | – |
| Total |  | 10-0.4 | 17-3.1 | 24-8.7 | 69-46.5 | 68-55.4 | 103-98.0 | 64-64.8 | a-35.4 | a-40.6 |

a) Estimate on the number of new plants not available.

*Sources:* *Summary of Nuclear Power and Fuel Cycle Data*, OECD Nuclear Energy Agency, 1985;
World List of Nuclear Power Plants, Nuclear News, February 1985;
*Nuclear Power Reactors in the World*, International Atomic Energy Agency, Vienna, 1985

# COSTS ESTIMATES FOR DECOMMISSIONING OF POWER REACTORS IN DIFFERENT COUNTRIES

## Table 12
### DECOMMISSIONING COSTS IN THE UNITED STATES[a], CONTINGENCY INCLUDED AT 25 %
Millions of January 1984 U.S. dollars

**1 175 MWe PWR**

|  | Pre-decommissioning Engineering | Decommissioning Operations | Rad Waste Transport and Disposal | Total Cost |
|---|---|---|---|---|
| Stage 1 | 2 | 20 | 2 | 24 |
| Stage 2 | 4 | 48 | 4[b] | 56 |
| Stage 3 Immediately | 5 | 55 | 28 | 88 |
| Stage 1 + 30 years Storage + Stage 3 after 30 years | 5 | 76 | 28 | 109 |
| Storage after Stage 1 for 30 years | 0 | 8 | 0 | 8 |
| Stage 3 after 30 years | 4 | 48 | 26 | 78 |
| Stage 2 + 100 years Storage + Stage 3 after 100 years | 7 | 122 | 14 | 143 |
| Storage after Stage 2 for 100 years | 0 | 27 | 0 | 27 |
| Stage 3 after 100 years | 4 | 47 | 10 | 61 |

**1 155 MWe BWR**

|  | Pre-decommissioning Engineering | Decommissioning Operations | Rad Waste Transport and Disposal | Total Cost |
|---|---|---|---|---|
| Stage 1 | 2 | 25 | 2 | 29 |
| Stage 2 | 4 | 60 | 4[b] | 68 |
| Stage 3 Immediately | 4 | 68 | 28 | 100 |
| Stage 1 + 30 years Storage + Stage 3 after 30 years. | 5 | 92 | 28 | 125 |
| Storage after Stage 1 for 30 years | 0 | 8 | 0 | 8 |
| Stage 3 after 30 years | 4 | 60 | 26 | 90 |
| Stage 2 + 100 years Storage + Stage 3 after 100 years | 7 | 145 | 13 | 165 |
| Storage after Stage 2 for 100 years | 0 | 27 | 0 | 27 |
| Stage 3 after de 100 years | 4 | 59 | 10 | 73 |

a) The estimates are based on NUREG/CR-0130 and NUREG/CR-0672. The costs were escalated to reflect general inflation in the U.S. from January 1978 to January 1984. Additional adjustments were made in some cost elements, such as waste transportation and disposal, to reflect current conditions; other costs which were added include changes to predecommissioning engineering, contractor management and on-site mobilisation and demobilisation.
b) Most of rad waste stored in reactor containment.

Table 13
## DECOMMISSIONING COSTS IN THE FEDERAL REPUBLIC OF GERMANY[a]
Millions of January 1984 U.S. dollars

### 1 200 MWe PWR

|  | Pre-decommissioning Engineering | Decommissioning Operations | Rad Waste Transport and Disposal | Contingency 20 % | Total Cost |
|---|---|---|---|---|---|
| Stage 1 | 2 | 7 | 1 | 2 | 12 |
| Stage 3 Immediately | 8 | 59 | 21 | 16 | 96 |
| Stage 1 + 30 years  Storage + Stage 3 after 30 years | 9 | 66 | 14 | 18 | 107 |
| Storage after Stage 1 for 30 years | 0 | 1 | 0 | 0 | 1 |
| Stage 3 after 30 years | 7 | 58 | 13 | 16 | 94 |

### 800 MWe BWR

|  | Pre-decommissioning Engineering | Decommissioning Operations | Rad Waste Transport and Disposal | Contingency 20 % | Total Cost |
|---|---|---|---|---|---|
| Stage 1 | 1 | 7 | 1 | 3 | 12 |
| Stage 3 Immediately | 8 | 58 | 19 | 30 | 115 |
| Stage 1 + 30 years  Storage + Stage 3 after 30 years | 8 | 68 | 13 | 32 | 121 |
| Storage after Stage 1 for 30 years | 0 | 1 | 0 | 0 | 1 |
| Stage 3 after 30 years | 7 | 60 | 12 | 28 | 107 |

a)  The estimates are based on data in Watzel et al., *Decommissioning of nuclear power stations in the Federal Republic of Germany at the end of their service life*, Progress report of the Association of German Engineers, Environmental Technology, Series 15, No. 18, 1982.

Table 14
## DECOMMISSIONING COSTS IN SWEDEN[a]
## CONTINGENCY INCLUDED AT 25%

| Reactor Type | Net power (MWe) | Costs for Stage 3 immediately, January 1984 U.S. dollars | | |
|---|---|---|---|---|
|  |  | Decommissioning operations | Waste transport and disposal | Total |
| BWR | 440 | 45 | 10 | 55 |
| BWR | 570 | 60 | 10 | 70 |
| BWR | 750 | 70 | 10 | 80 |
| BWR | 900 | 90 | 15 | 105 |
| BWR | 1 050 | 100 | 15 | 115 |
| PWR | 915 | 70 | 10 | 80 |

a)  Estimations are based on the report *Technology and Costs of Dismantling a Swedish Nuclear Power Plant*, SKBF/KBS TR 79-22, Swedish Nuclear Fuel and Waste Management Co., 1979.

## Table 15
### DECOMMISSIONING COSTS IN FINLAND[a]
January 1984 U.S. dollars

| | Pre-decommissioning Engineering | Decommissioning Operations | Storage and Surveillance | Waste Management | Contingency | Total |
|---|---|---|---|---|---|---|
| Stage 3 for 2 × 440 MWe | 2 | 42 | – | 13 | 17 | 74 |
| Stage 1 followed by 31-year storage and Stage 3 for 2 × 710 MWe BWR | 3 | 80 | 8 | 19 | 22 | 132 |

a) Original estimates given in 1985 Finnish Marks (FIM) have been changed to 1984 Finnish Marks assuming 6.1 per cent inflation during 1984 and then converted to U.S. dollars using the rate 1 U.S. dollar = 5.81 FIM. In these estimates, dismantling is restricted to contaminated parts (site continues to be used for power plant purposes).

## Table 16
### DECOMMISSIONING COSTS IN CANADA FOR 4 × 515 MW(e) HWR[a]
January 1984 U.S. dollars

| | Pre-decommissioning Engineering and Decommissioning Operations | Waste Transport and Disposal | Contingency (10 %) | Total |
|---|---|---|---|---|
| Stage 1 | 14 | 1 | 1 | 16 |
| Stage 3 immediately | 129 | 64 | 19 | 212 |
| Stage 1/30 years Storage/Stage 3 | 102 | 54 | 16 | 172 |
| Storage for 30 years | | 6 | 1 | 7 |

a) Estimates are based on Ontario Hydro reports 81156 (Preliminary Nuclear Decommissioning Cost Study, April 1981) and 82208 (Decommissioning by Immediate Dismantlement, Preliminary Cost Estimate for Pickering Nuclear Generating Station A, July 1982). The original estimate has been changed to U.S. dollars using the rate 1 U.S. dollar = 1.244 Can. dollars.

# DECOMMISSIONING COST DERIVATION

## The Concept of Discounted Future Costs and Levelised Costs

The present value "B" of a future amount "A" is:

$$B = A \frac{1}{(1 + i)^T} \tag{1}$$

where "T" is the time in years and "i" is the nominal interest rate which can reasonably be expected to be paid on invested funds over that time interval. In other words, if an amount "A" is to be available for decommissioning a plant "T" years in the future, it is necessary to invest only "B" today at interest rate "i".

If amount "A" must be increased to cover rising costs due to inflation at an annual rate of "a" over the "T" years, then the amount "B" to be invested today must be increased to:

$$B = A \frac{(1 + a)^T}{(1 + i)^T} \tag{2}$$

Both inflation rates and attainable interest rates can vary significantly over time and reasonable values for "i" and "a" can be difficult to forecast with any accuracy. However, as can be seen in equation 2, their absolute values are less important than their ratio. In fact a value "r" can be defined as the *real* interest rate, that is the interest rate above inflation, such that:

$$(1 + r) = \frac{(1 + i)}{(1 + a)}$$

so that:

$$B = \frac{A}{(1 + r)^T} \tag{3}$$

The attainable real interest rate can vary in the short-term but tends to be more stable over long periods. Its value will depend on the type of investment but for utility investments most OECD countries consider a value of about 0.05 (i.e. 5 per cent) to be appropriate for present-worth calculations. A value for "r" (usually called the discount rate in present-worth calculations) of 5 per cent has been used for reference case calculations in this and other NEA studies (1, 2), and sensitivity analyses have considered the range from 0 per cent to 10 per cent.

To make future costs and revenues comparable with each other, both must be discounted to determine their present value at some reference date. If it is decided to obtain revenue from the sale of electricity during the plant's operating lifetime to pay for decommissioning, then the total present worth of the revenues received for this purpose during the operating lifetime must equal the present worth of the decommissioning costs, all discounted to the same reference time. For this study the reference time is the date when the plant starts commercial operation.

The constant fee which must be charged per kilowatthour of electricity generated to make the two present-worthed values exactly balance is called the *levelised cost*. More detailed discussion on levelised cost calculations can be found in Refs. 1 and 2.

## The Concept of Constant Money and Requirements for Comparisons With Other Studies

The costs of decommissioning are estimated, in this report, in 1984 U.S. dollars. That is, if the plant had been decommissioned in 1984 it is believed that the job could have been done for the amounts shown. If a plant were to be decommissioned in the year 2020 (or any future year) with the same methods, material and labour, and had there been no inflation in the interval, it would still be expected to cost the same amount of money. Inflation inevitably would have raised the costs (or, it might be said, decreased the value of the dollar) but it is difficult to know in advance what the inflation rate will be over this long period. Therefore, it is easier to continue to quote the costs in *constant 1984 money*. When decommissioning starts, the cost in 2020 dollars could reasonably be expected to be equal to the 1984 costs multiplied by the composite inflation index for the 1984-2020 period.

This approach does not neglect inflation. It is allowed for in discounting future costs at the *real interest rate* instead of the higher *nominal interest rate*, as explained previously.

The constant money approach allows the cost of decommissioning of plants to be considered independent of the year when the decommissioning starts. However, when comparisons are made between decommissioning costs and other costs of electricity generation, the same year's dollar values must be used. The generating costs used for comparison in this report are taken from another NEA study (2), which also uses constant 1984 U.S. dollars.

Other ground rules must also be the same for comparisons to be valid and these two studies have therefore each used a 5 per cent discount rate and a 25-year plant life for their reference cases and costs are discounted to the start of commercial operation. The generating costs were specifically based on plants expected to start operating in mid-1995. Because of the constant money approach the costs are not very sensitive to this date but some allowance for real price increases (i.e. cost increases above the rate of inflation) have been made. Thus, the use of decommissioning cost estimates for currently operating plants is not strictly consistent with the assumptions used for other costs. However, within the accuracy of the estimates available, this does not affect the conclusions.

**Decommissioning Cost Present Value Calculation**

Figure 3 depicts the events of a Stage 1 or 2 decommissioning followed by a storage period prior to Stage 3 removal. The time "T" in the above equation (3) is thus replaced by the appropriate intervals:

L = duration of power generation,

k = time from reactor shutdown to start of decommissioning, assumed to be one year,

$m_1$ = time from start of decommissioning to year of peak expenditure, assumed to be 1.5 and 2.5 years for Stage 1 and Stage 2, respectively,

$m_2$ = time from year of peak expenditure to completion of given Stage, 2.5 years,

$m_3$ = storage time, assumed to be 30 years and 100 years for Stage 1 followed by Stage 3, or for Stage 2 followed by Stage 3, respectively,

$m_4$ = time from end of storage to year of peak expenditure, assumed to be 1.5 years for both cases.

If, for example, $d_1$ and $d_3$ are the total costs for Stages 1 and 3, respectively and $d_2$ is the cost per year for storage, then the present value, D, of the total decommissioning effort for Stage 1 followed by Stage 3 will be :

$$D = d_1 \frac{1}{(1 + r)^{L + k + m_1}} + S + d_3 \frac{1}{(1 + r)^{L + k + m_1 + m_2 + m_3 + m_4}} \tag{4}$$

where

$$S = \frac{d_2}{r} \left(1 - \frac{1}{(1 + r)^{m_3}}\right) \cdot \frac{1}{(1 + r)^{L + k + m_1 + m_2}} \tag{5}$$

Equation 4 makes the approximation that all costs for a given stage occur at a single point in time, the year of peak expenditure for that stage, rather than separately discounting each year's expenditure. However, the error introduced is trivial.

The equation holds for all three decommissioning strategies considered, though for immediate Stage 3 the first two terms would be zero, as would exponents $m_2$, $m_3$ and $m_4$ in the third term.

**Levelised Cost Calculations**

If decommissioning is funded by a fee charged to each unit of electricity produced by the reactor over its lifetime, then the present value of the revenue stream, discounted to the same reference time as for decommissioning costs, must be equal to D given by equation 4. Therefore:

$$D = C \times P \times H \tag{6}$$

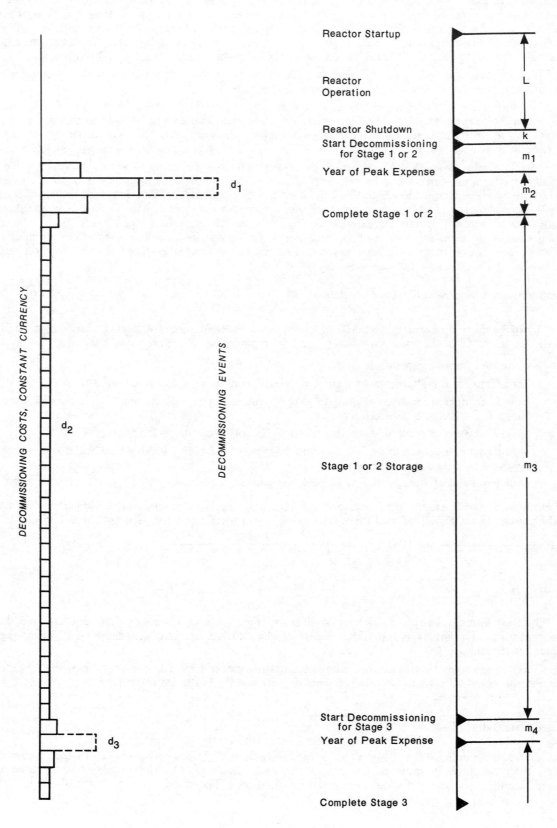

60

where

$\quad$ C $\quad$ = $\quad$ levelised decommissioning charge,
$\quad$ P $\quad$ = $\quad$ reactor electric power,
$\quad$ H $\quad$ = $\quad$ discounted number of full-power hours in the plant's operating lifetime.

If $f_t$ is the plant's load factor in the year t, H can be calculated as:

$$H = 8\,760 \sum_{t=1}^{L} f_t/(1 + r)^{t-1/2}$$

or

$$H = 8\,760 \times F \sum_{t=1}^{L} 1/(1 + r)^{t-1/2}$$

where

$$F = \sum_{t=1}^{L} f_t/(1 + r)^{t-1/2} / \sum_{t=1}^{L} 1/(1 + r)^{t-1/2}$$

is called the average levelised load factor.

In another NEA study (2) total levelised generating costs were developed for the same reactor lifetime scenario as used in this report for the reference case, namely 25 years with an average levelised load factor of about 72 per cent (4 000 hours in the first year, 5 000 hours in the second year and 6 600 hours per year thereafter). Two variants were also considered, the first assuming the same hours per year as the reference case but a 20-year operating lifetime (giving a levelised load factor of about 71 per cent), the second assuming a 25-year lifetime but only 6 000 operating hours per year at equilibrium (load factor about 66 per cent). In addition to reference discount rate, 5 per cent, discount rates of 0 and 10 per cent were also evaluated. For this study, an additional variant was defined (Variant 3): it assumes the same operating hours per year as the reference scenario but a 30-year reactor lifetime. The values of H for each of these cases described are as follows:

| Discount Rate | Reference case | Variant 1 | Variant 2 | Variant 3 |
|---|---|---|---|---|
| 0 % | 160 800 | 126 800 | 147 000 | 193 800 |
| 5 % | 91 293 | 79 300 | 83 771 | 99 940 |
| 10 % | 58 967 | 54 100 | 54 347 | 61 389 |

Equation (6) can be rewritten as:

$$C = \frac{D}{P \times H} \tag{7}$$

to obtain the levelised decommissioning cost when total decommissioning cost and total lifetime electricity production is known. If P is given in kilowatts and D in dollars, C will be given in $/kWh and must be multiplied by 1 000 to arrive at costs in mills/kWh (1 000 mills = 1 dollar).

### Sample Calculation

$\quad$ – $\quad$ Country $\quad$ : $\quad$ U.S.A.:
$\quad$ – $\quad$ Reactor $\quad$ : $\quad$ 1 300 MW(e) PWR;
$\quad$ – $\quad$ Strategy $\quad$ : $\quad$ Stage 1 plus 30 years storage plus Stage 3;
$\quad$ – $\quad$ Case $\quad$ : $\quad$ Reference (5 per cent discount rate: 25-year lifetime)
$\quad$ – $\quad$ Variables $\quad$ : $\quad$ (see Figure 3 for definitions)

$L$ $\quad$ = $\quad$ 25 years;
$K$ $\quad$ = $\quad$ 1 year; $\qquad$ 27.5 years
$m_1$ $\quad$ = $\quad$ 1.5 years;
$m_2$ $\quad$ = $\quad$ 2.5 years; $\qquad$ 61.5 years
$m_3$ $\quad$ = $\quad$ 30 years;
$m_4$ $\quad$ = $\quad$ 1.5 years.

$d_1$ $\quad$ = $\quad$ $26 million $\qquad$ [See Appendix 2; estimates modified to correspond to a
$d_2$ $\quad$ = $\quad$ $0.3 million/year $\qquad$ 1 300 MW(e) plant]
$d_3$ $\quad$ = $\quad$ $86 million

From equations 4 and 5:

$$D = \frac{26 \times 10^6}{(1.05)^{27.5}} + \frac{0.3 \times 10^6}{0.05} \left[1 - \frac{1}{1.05^{30}}\right] \times \frac{1}{1.05^{30}} + \frac{86 \times 10^6}{1.05^{61.5}}$$

    = \$12.14 million = total decommissioning cost in 1984 dollars discounted to start of commercial operation.

[Note:    $D = 12.14 \times 10^6 \times (1.05)^{25} = \$41$ million 1984 dollars discounted to time of plant shutdown, as shown in Table 8.]

From equation 7, with H = 91 293 for the reference case,

$$C = \frac{12.14 \times 10^6 \text{ \$}}{91\ 293 \times 1.3 \times 10^6 \text{ kWh}} = 0.10 \text{ mills/kWh}$$

    The total levelised cost of power production for the U.S. PWR in the same case (2) is 43.8 mills/kWh. Thus decommissioning would account for 0.2 per cent of the levelised cost of electricity from the PWR.

### Summary of Levelised Decommissioning Costs

    A summary of the levelised cost calculations on the basis of the countries' estimates for decommissioning costs is presented in Tables 12 and 13. Table 12 corresponds to the reference case and includes calculations for 0, 5, and 10 per cent discount rates. Table 13 gives the results for Variants 1 and 3 at 5 per cent discount rate. Variant 2 has not been considered separately; in comparison with the reference case, the cost effect of lower load factor at equilibrium is inversely proportional to the change in levelised average load factor (Equation 7).

## REFERENCES

1. Economics of the Nuclear Fuel Cycle, OECD Nuclear Energy Agency, 1985.
2. Projected Costs of Generating Electricity from Nuclear and Coal-Fired Power Stations for Commissioning in 1995, OECD Nuclear Energy Agency, 1986.

Table 17

**DECOMMISSIONING COSTS PER UNIT OF ELECTRICITY PRODUCED**

**25-YEAR UNIT LIFETIME**

January 1984 U.S. mills/kWh

| Country | Reactor type | Stage 3 immediately | | | Stage 1/30 years/ Stage 3 | | | Stage 2/100 years/ Stage 3 | | |
|---|---|---|---|---|---|---|---|---|---|---|
| | | Discount rate, % | | | | | | | | |
| | | 0 | 5 | 10 | 0 | 5 | 10 | 0 | 5 | 10 |
| United States | PWR | 0.6 | 0.2 | 0.1 | 0.7 | 0.1 | 0.0 | 1.0 | 0.1 | 0.1 |
| | BWR | 0.7 | 0.2 | 0.1 | 0.9 | 0.1 | 0.0 | 1.1 | 0.2 | 0.1 |
| Federal Republic | PWR | 0.7 | 0.3 | 0.1 | 0.7 | 0.1 | 0.0 | | | |
| of Germany | BWR | 1.0 | 0.4 | 0.2 | 1.1 | 0.1 | 0.0 | | | |
| Sweden | PWR | 0.7 | 0.2 | 0.1 | | | | | | |
| | BWR | 1.0 | 0.4 | 0.2 | | | | | | |
| Finland | PWR | 0.6 | 0.2 | 0.1 | | | | | | |
| | BWR | | | | 0.7 | 0.1 | 0.0 | | | |
| Canada | HWR | 0.9 | 0.3 | 0.1 | 0.7 | 0.1 | 0.0 | | | |

## Table 18
### DECOMMISSIONING COSTS PER UNIT OF ELECTRICITY PRODUCED FOR TWO LIFETIME VARIANTS
### 5 PER CENT DISCOUNT RATE
#### January 1984 U.S. mills/kWh

| Country | Reactor type | Stage 3 immediately | | Stage 1/30 years/ Stage 3 | | Stage 2/100 years/ Stage 3 | |
|---|---|---|---|---|---|---|---|
| | | Unit Lifetime, Years | | | | | |
| | | 20 | 30 | 20 | 30 | 20 | 30 |
| United States | PWR | 0.3 | 0.2 | 0.2 | 0.1 | 0.2 | 0.1 |
| | BWR | 0.4 | 0.2 | 0.2 | 0.1 | 0.2 | 0.1 |
| Federal Republic of Germany | PWR | 0.4 | 0.2 | 0.1 | 0.1 | | |
| | BWR | 0.6 | 0.3 | 0.2 | 0.1 | | |
| Sweden | PWR | 0.4 | 0.2 | | | | |
| | BWR | 0.5 | 0.2 | | | | |
| Finland | PWR | 0.3 | 0.2 | | | | |
| | BWR | | | 0.1 | 0.1 | | |
| Canada | HWR | 0.5 | 0.2 | 0.1 | 0.1 | | |

*Appendix 4*

# EXCHANGE RATES

## EXCHANGE RATE PER US$ AT 1.1.1984

| Country | Currency | Exchange Rate |
|---|---|---|
| Australia | Dollar | 1.1205 |
| Austria | Schilling | 19.341 |
| Belgium | Franc | 55.64 |
| Canada | Dollar | 1.2444 |
| Denmark | Krone | 9.8750 |
| Finland | Markka | 5.8100 |
| France | Franc | 8.3475 |
| Federal Republic of Germany | DM | 2.7238 |
| Greece | Drachma | 98.670 |
| Ireland | Pound | 0.8811 |
| Italy | Lire | 1659.5 |
| Japan | Yen | 232.20 |
| Netherlands | Gulden | 3.0645 |
| New-Zealand | Dollar | 1.5279 |
| Norway | Krone | 7.7425 |
| Portugal | Escudo | 131.65 |
| Spain | Peseta | 156.70 |
| Sweden | Krona | 8.0010 |
| Switzerland | Franc | 2.1795 |
| Turkey | Lira | 282.80 |
| United Kingdom | Pound | 0.6894 |
| United States | Dollar | 1 |

Sources: *Main Economic Indicators*, OECD, February 1984.
*Bulletin of the European Communities*, 16, 12 (1983).

*Appendix 5*

# MEMBERS OF THE WORKING GROUP ON DECOMMISSIONING

| | | |
|---|---|---|
| *Belgium* | Mr. P. Dozinel | Tractionel |
| | Mr. J.C. Moureau | Ministère de la Santé Publique |
| | Mr. M. Preat | Tractionel |
| *Canada* | Mr. R. Bennett | Atomic Energy of Canada Ltd. |
| | Mr. P. De | Atomic Energy of Canada Ltd. |
| | Mr. H.B. Merlin | Energy, Mines and Resources |
| | Mr. J. Saroudis | Atomic Energy of Canada Ltd. |
| *Finland* | Mr. T. Kukkola | Imatran Voima Oy |
| *France* | Mr. F. Anselin | Commissariat à l'Energie Atomique |
| | Mr. A. L. Crégut | Commissariat à l'Energie Atomique |
| *Germany, Federal Republic of* | Mr. R. Görtz | Brenk Systemplanung |
| | Mr. W. Zimmermann | Kernforschungszentrum Karlsruhe |
| *Italy* | Mr. M. Conti | Ente Nazionale Energia Alternative |
| | Mr. T. Vitiello | Ente Nazionale per l'Energia Elettrica |
| *Japan* | Mr. T. Egashira | Japan Atomic Energy Research Institute |
| | Mr. T. Imamura | Japanese Delegation to the OECD |
| *Spain* | Mr. R. San Martin | Ministerio de Industria |
| *Sweden* | Mr. H. Forsström | Swedish Nuclear Fuel and Waste Management Co. |
| *United Kingdom* | Mr. A. Colquhoun | British Nuclear Fuel plc |
| | Mr. A.R. Gregory | Central Electricity Generating Board |
| | Mr. J.D. Hart | Central Electricity Generating Board |
| | Dr. S.M. Stearn | Department of Environment |
| *United States* | Mr. E.G. DeLaney (Chairman) | Department of Energy |
| | Ms. J. Mickelson (Consultant) | UNC Nuclear Industries |
| | Mr. A. Reynolds | United States Delegation to the OECD |
| *CEC* | Mr. R. Bisci | Programme Déclassement |

| | | |
|---|---|---|
| *IAEA* | Mr. D. E. Saire | Waste Management Section |
| *OECD/NEA* | Mr. T. Sasaki | Nuclear Science and Development |
| | Mr. K. Bragg | Radiological Protection and Waste Management Division |
| | Mr. M. Crijns | Nuclear Development Division |
| | Ms. M. Dawson | Nuclear Development Division |
| | Mr. O. Ilari | Radiological Protection and Waste Management Division |
| | Mr. W.T. Potter | Nuclear Development Division |
| | Dr. P. Silvennoinen | Nuclear Development Division |
| | Mr. H.E. Thexton | Nuclear Development Division |
| | Mr. J. Vira (Secretary) | Nuclear Development Division |

# COUNTRY SPECIFIC ANNEXES

The information presented below comes primarily from answers to a questionnaire distributed to NEA Member States. The questionnaire asked for the following types of information:

1. National policies and strategies for decommissioning, radioactive waste handling and disposal and optimisation of decommissioning plans;

2. Social considerations taken into account in forming decommissioning policies or strategy;

3. Decommissioning regulations;

4. Major decommissioning projects that are scheduled for the next 10 years or longer;

5. Facility repair and maintenance work that can be related to decommissioning;

6. Demonstration projects that are needed, but not yet planned, that can provide additional technical experience and cost data;

7. Any projects planned or underway that will futher develop and demonstrate decommissioning-related technology;

8. Costs and manhours for past and planned major decommissioning projects;

9. Scope of work used for performing total decommissioning for each past and planned major project;

10. Principal alternative methods of financing decommissioning.

The following annexes do not contain information on every topic listed above. As would be expected, most of the respondents reported they had no experience or plans in several areas; furthermore, some of the material provided in response to the questionnaire is reported in other sections of this document.

# BELGIUM

## NATIONAL POLICY

*Waste Management*:   Reprocess spent fuel; vitrify HLW, ILW and TRU wastes in a geologic repository; dispose of LLW in the ocean, if possible, or by shallow land burial. Belgium has a major study underway to evaluate the safety of disposing of HLW and spent fuel in a clay formation. The management of radioactive waste is the responsibility of a public body, the National Organisation for Radioactive Waste and Fissile Materials (ONDRAF), which is competent in the fields of spent fuel and waste transport and storage, waste conditioning and waste disposal.

*Decommissioning Strategy*:   Not yet developed.

*Financing Provisions for Decommissioning*:   Each electricity producing company is required to establish an internal fund for nuclear plant decommissioning. Funds are raised through annual contributions during plant lifetime (conventionally assumed as 20 years). Together with the interest accrued these contributions must, in 30 years from plant startup, make 12 per cent of the investment cost (excluding interest during construction) presently needed to build such a plant. Interest calculation is based on rates customarily used for present worth calculations. Annual contributions to the fund are taken into account in the kWh prices for electricity.

*Social Considerations*:   No specific application to questions relating to the decommissioning of nuclear installations.

## REGULATIONS

No specific legislation concerning decommissioning. Such activities are covered by the general regulations governing nuclear activities.

## DEVELOPMENT AND DEMONSTRATION OF RELATED TECHNOLOGY

1.     *Eurochemic reprocessing plant* – the plant may be refurbished and operated again, or it may be decommissioned totally; either activity should provide useful experience and information concerning decommissioning costs for nuclear facilities that use chemical processes.

2.     *Nuclear Energy Research Centre (Mol)* – development of various techniques that might be used in treating decommissioning wastes: high-temperature incineration; cutting and compaction of solid wastes; conditioning of concentrates by bituminisation, concreting or encapsulation in resins. Various Belgian organisations also collaborate with the European Communities under their decommissioning programs – specific Belgian contributions include post-accident decontamination studies at TMI in the U.S., development of a mobile robot able to inspect, control and repair a reactor in operation and development of large containers for the transport of wastes from decommissioning PWRs.

# CANADA

## NATIONAL POLICY

*Waste Management*:   Prepare for deep geologic disposal of either spent fuel or immobilised HLW and TRU wastes; store LLW/ILW until disposal facilities matched to the intrinsic properties of the various types of wastes are ready. Development of waste management technology is handled principally by Eldorado Resources Limited, Atomic Energy of Canada Limited (AECL) and Ontario Hydro.

*Decommissioning Strategy*:   No specific strategy delineated; decommissioning regarded as simply another stage in a nuclear facility's life cycle, subject to virtually the same policies as those applied to siting, construction and operation.

*Financial Provisions for Decommissioning*: Facility owners/operators are responsible for decommissioning of their facilities and are free to accomplish this by whatever mode they wish to employ, provided they comply with the pertinent national nuclear objectives, requirements and performance standards as established by the Atomic Energy Control Board (AECB). The method currently employed for financing decommissioning projects follows the principal that when the costs of removal of fixed assets can be reasonably estimated and are significant, the amounts should be charged to operations over the remaining service life of the asset.

*Social Considerations*: The AECB endeavours to obtain, and take into account at an early stage, public opinion concerning nuclear regulatory policies, requirements and guidelines. This approach will be followed in the case of decommissioning. Two important concerns are the choice of decommissioning mode and the disposal of decommissioning wastes.

## REGULATIONS

There are currently no regulations specific to decommissioning; decommissioning activities are generally subject to the Atomic Energy Control Regulations SOR/74-334, 1974, as subsequently amended. The proposed general amendments to the regulations issued as Consultative Document C-83 in April 1986, contain requirements concerning the issuance of decommissioning licences.

## DEVELOPMENT AND DEMONSTRATION OF RELATED TECHNOLOGY

1.      Search for decontamination techniques that reduce the volumes and acid and complexant content of decontamination wastes generated.

2.      Development of equipment for the remote removal and replacement of contaminated equipment (remotely controlled plasma arc cutting and TIG welding have been used in re-tubing CANDU reactor vessels).

3.      Work in improved waste conditioning techniques: controlled-air incineration; bituminisation of incinerator ash; shredding and bituminisation of contaminated *trash*; use of cement and polyester resin as immobilisation agents.

4.      Development of improved waste containers.

## ANALYSIS OF DECOMMISSIONING COSTS

1.      Percentage cost breakdown for Gentilly-1 Decommissioning Programme (actual cost data are not available outside the AECL):

| Activity | Stage 3 Decommissioning immediately after final shutdown | Stage 1 Decommissioning immediately after final shutdown; Stage 3 Decommissiong after 30 years |
|---|---|---|
| Engineering and planning | 7.5 | 5.9 |
| Administration and on-site construction management | 21.3 | 13.8 |
| Actual decommissioning | 54.7 | 69.6 |
| Contingency | 16.5 | 10.7 |

2.      Percentage cost breakdown for major tasks for the case of Stage 3 decommissioning Gentilly-1 immediately after shutdown and extending over a six year period.

| Activity | Reactor building | Service building | Turbine building | Pumphouse and switch yard | Total |
|---|---|---|---|---|---|
| Decontamination | 1.0 | – | 0.4 | – | 1.4 |
| Dismantling | 9.1 | 3.8 | 6.9 | 0.6 | 20.4 |
| Packaging | 3.5 | 0.1 | 1.9 | 0.1 | 5.6 |
| Transportation | 1.4 | – | – | – | 1.4 |
| Total | 15.0 | 3.9 | 9.2 | 0.7 | 28.8 |

# FINLAND

## NATIONAL POLICY

*Waste Management*:   Prepare for final disposal of spent fuel in a geologic repository in case contractual arrangement for transferring spent fuel or reprocessing waste irrevocably outside the country cannot be achieved. Reactor wastes are packaged and disposed of in a rock cavern at the plant site.

*Decommissioning Strategy*:   Go to Stage 3 as soon as practical; the plant site and some structures are to be reused for energy production.

*Financial Provisions for Decommissioning*:   The owner is responsible for taking care of the plant and its wastes in a reasonable period of time. At present, the utilities are obliged to set aside funds each year for decommissioning; long-term control and use of the decommissioning fund is to be stipulated in the new nuclear energy law.

## REGULATIONS

Regulations specific to decommissioning have not been formulated, so the general provisions for nuclear work apply. No criteria have been set officially for licensing termination or unrestricted release, but the following criteria are used in design:

| | |
|---|---|
| Surface $\alpha$ | $3.7 \times 10^3$ Bq/m$^2$ |
| Surface ß, $\gamma$ | $3.7 \times 10^4$ Bq/m$^2$ |
| Specific Activity | 74 Bq/g |

## DEVELOPMENT AND DEMONSTRATION OF RELATED TECHNOLOGY

No major activities under way at present.

## ESTIMATED DECOMMISSIONING COSTS

Costs and manpower requirements have been estimated for decommissioning the Loviisa and Olkiluoto Nuclear Power Stations. The breakdown of the estimates is as follows (costs in millions of 1984 U.S. dollars):

| | Loviisa 1 & 2 | | Olkiluoto 1 & 2 | |
|---|---|---|---|---|
| Project Management and Design | 560 m mo | 2.4 | 720 m mo | 3.3 |
| Safe Storing | | | | 8.3 |
|    Storing Costs | | | 1 600 m mo | |
|    Surveillance (31 years) | | | 1 000 m mo | |
| Dismantling | | | | 71.1 |
|    Decontamination | | 0.5 | | |
|    RPV and Internals | 90 m mo | 2.7 | 1 000 m mo | |
|    Contaminated Systems | 9 510 m mo | 29.6 | 16 700 m mo | |
|    Contaminated Construction | 240 m mo | 1.0 | 260 m mo | |
| Waste Management | | 13.1 | | 19.2 |
| Operation | | 8.3 | | 8.6 |
| Contingency | | 17.3 | | 22.1 |
| Total | | 74.9 | | 132.6 |

# FRANCE

## NATIONAL POLICY

*Waste Management*:

- For short-lived radionuclides the wastes are isolated by suitable embedding for 300 years and sent to a controlled shallow land burial site. Beyond 300 years the radioactivity will be negligible.
- For long-lived radionuclides and fission products the wastes are isolated or vitrified and buried in deep geological formations.
- Liquid wastes are processed by conventional techniques such as decontamination and evaporation.

*Decommissioning Strategy*:   Today decommissioning is considered to be at an experimental stage. Dismantling operations are examined on a case by case basis, followed by pilot operations and research and development of suitable technologies.

*Financial Provisions for Decommissioning*:   The utility takes into account the decommissioning costs in the same way as the construction costs. For the Atomic Energy Commission the annual funding allows dismantling of surplus facilities.

*Social Considerations*:   Social considerations must be taken into account but are not expected to provide significant impact on future decommissioning activities.

## REGULATIONS

Actual regulations for nuclear facilities are suitable for decommissioning. Specific regulations will be formulated on the basis of the decommissioning experience. Activity limits for free release must be carefully set up.

## DEVELOPMENT AND DEMONSTRATION OF RELATED TECHNOLOGY

- Decommissioning operations are managed in order to obtain experience and induce Research and Development.
- R and D focusses on the following items:
  1. Decommissioning safety assessment;
  2. Robotics – Remote handling;
  3. Cutting tools and techniques;
  4. Decontamination means and processes;
  5. Decommissioning waste treatment;
  6. Methodology for general studies and cost analyses.

## ESTIMATED DECOMMISSIONING COSTS

Computerised cost analysis systems are set up to evaluate costs and take into account the experimental input.

# FEDERAL REPUBLIC OF GERMANY

## NATIONAL POLICY

***Waste Management***:   Build and operate AFR storage facilities for spent fuel; reprocess and vitrify HLW; prepare for disposal of all wastes, potentially including spent fuel, in geologic repositories. Reprocessing wastes are to be placed in a salt dome repository, while decommissioning and reactor wastes are to go into the Konrad Mine (crystalline rock) repository, which is expected to be ready in 1989. The private sector is responsible for radioactive waste conditioning and storage; the government for waste disposal.

***Decommissioning Strategy***:   No generally preferable approach to decommissioning has been identified. Licensing is based on the same legal provisions and guiding principles as in other fields of nuclear technology. The development of specific rules and guildelines shall proceed according to the needs observed in decommissioning experience.

***Development and Demonstration of Related Technology***:   Though the general feasibility of decommissioning is well established, progress in certain areas is desirable. Therefore, the government of the Federal Republic of Germany has sponsored research and development efforts to improve decommissioning techniques covering, but not being restricted to, explosive dismantling of reinforced concrete, piping and pressure vessels after embrittling by cooling with liquid nitrogen, powder and plasma cutting, decontamination and melting of radioactive metals.

Furthermore, as a member of the European Community, the Federal Republic of Germany has taken part in the extensive programme of the Community on the decommissioning of nuclear power plants.

***Estimated Decommissioning Costs***:   See Table 13 of Appendix 2 in this report.

***Financial Provisions for Decommissioning***:   The utilities are responsible for all decommissioning actions including cost. They make provision for liabilities and charges on a voluntary basis based on estimated cost and facility operation time.

***Social Considerations***:   Questions about the feasibility of decommissioning, raised by the anti-nuclear movement in Germany, provided part of the incentive for the KKN decommissioning project. At present, social considerations are not expected to provide significant impact on future decommissioning activities.

## REGULATIONS

1.      A licence is required for the decommissioning of an installation, the safe enclosure of a finally decommissioned installation or the dismantling of the installation or parts thereof – Atomic Energy Act of October 31, 1986, Annex 2.

2.      The Atomic Energy Act and the ordinances based on it are executed by the States (Lander) on behalf of the Federal Government. The state authorities are supervised by the Federal Minister of the Interior, who is advised by two expert groups, the Reactor Safety Commission (RSK) and the Radiological Protection Commission (SSK) – Grundgesetz, Article 87 c.

3.      Nuclear power plants are required to be in such a condition that they can be decommissioned in compliance with the Radiation Protection Regulations. A concept for the removal of the plant after its final shutdown is to be provided. Presentation of this concept is demanded in the licensing procedure for new nuclear power plants. Nuclear Power Plant Safety Criteria, promulgation of October 21, 1977, Criterion 2.10.

4.      Design of nuclear facilities is to make allowance for ultimate decommissioning of the plant, its security and/or its disposal. Components are to be designed and arranged in such a way that they can be decontaminated, disassembled and transferred inside the plant with as low a radiation exposure as possible. RSK guidelines for pressurised water reactors, 3rd Edition, October 14, 1981, Item 16.

5.   The following criteria have been estabished as a basis of licensing for release:

    *a)*   Surface Contamination    $0.037$ $Bq/cm^2$: $\alpha$ emitters
                                        $0.37$ $Bq/cm^2$: $\beta, \gamma$ emitters

    *b)*   Specific Activity          $10^{-4}$ allowance/g.

(*values of the allowance* are provided in the *Ordinance on Protection.*)

A more detailed guideline for release criteria is presently under consideration.

# ITALY

## NATIONAL POLICY

*Waste Management*:   Low radioactive waste handling and conditioning techniques have been developed and can now be considered commonly available. Suitable locations (shallow land burial or underground cavities for a centralised temporary storage) have been investigated. Actions are in progress to select the most appropriate sites.

*Decommissioning Strategy*:   As major activities have not been done so far, decommissioning is considered to be at an experimental stage. The policy on power plants decommissioning is to implement Stage 1 and possibly further steps towards Stage 2, then to delay implementation of Stage 3 up to 30 years. In the meanwhile appropriate procedures for total dismantling will be developed in the light of the current technology and suitable facilities for the disposal of the arising wastes will be made available.

*Financial Provision for Decommissioning*:   Italian operators of nuclear installations are essentially state-owned or state-controlled organisations, such as ENEL (State Electricity Board) and ENEA (National R&D Board for Nuclear and Other Alternative Energies). Therefore, no problem exists for the financing of decommissioning activities, other than suitable budgeting. Allowance for decommissioning costs is made in the evaluation of the unit electricity costs from projected nuclear power plants.

## REGULATIONS

All aspects concerning safety, radiation protection and environmental protection will fall in any case under appropriate national regulations. Specific reference to decommissioning activities will be included in the forthcoming revision of the Presidential Decree 13 February 1964 N.185, which is currently the basic law on nuclear safety and radiation protection. Meanwhile, regulatory provisions for the licensing activities are derived from applicable regulations in force.

## DEVELOPMENT AND DEMONSTRATION OF RELATED TECHNOLOGY

Some research activities are carried out to obtain experience mainly on techniques and equipment for decontamination and dismantlement. Projects are in progress for the decommissioning of important systems of one power plant.

## ESTIMATED DECOMMISSIONING COST

No real experience has been achieved on actual decommissioning cost for substantial installation, such as an industrial power plant.

# JAPAN

## NATIONAL POLICY

*Waste Management*:   Reprocess spent fuel; vitrify the HLW; dispose of alpha wastes and HLW packages in a deep geologic repository; dispose of LLW by ocean dumping, if possible, or by shallow land burial. Disposal of HLW and other long-lived wastes is a governmental responsibility, while treatment and disposal of low-level wastes are to be handled by the private sector. Decommissioning is to be carried out safely, in harmony with local society and with the intent to utilise the site efficiently afterwards.

*Decommissioning Strategy*:   Although decommissioning strategy is to be decided on a case by case basis, the principle is to dismantle and remove a reactor as soon as possible after its operation is terminated. In the case of commercial power reactors, the principle is to dismantle reactors after five to ten years safe storage.

*Financial Provisions for Decommissioning*:   In Summer 1985 the cost of decommissioning a 1 100 MW(e) commercial LWR on conditions mentioned above was estimated by a government-sponsored committee to be about 30 billion yen. The result of cost estimation may be reflected on electric charges, funding and taxation system.

*Social Considerations:*   Decommissioning has not yet been addressed specifically, although much effort goes into gaining public support for siting nuclear facilities.


## REGULATIONS


A report on decommissioning shall be submitted from applicant to the competent minister in advance and the competent minister may, if he deems necessary, designate the decommissioning method or establish requirements for decontamination, safety, etc. The competent ministers are the Minister of State for Science and Technology (experimental reactors), the Minister of International Trade and Industry (commercial power reactors) and the Minister of Transportation (nuclear ships) – Law for Regulation of Nuclear Source Material, Nuclear Fuel Material and Atomic Reactor, Article 38.

Licensing termination and unrestricted release criteria are not mentioned in the Japanese regulations.


## DEVELOPMENT AND DEMONSTRATION OF RELATED TECHNOLOGY

### Japan Atomic Energy Research Institute

The following techniques have been developed and will be demonstrated in the dismantlement of JPDR:
- A.   Estimation of radioactivity inventory;
- B.   Decontamination techniques for pipings and concrete structures;
- C.   Disassembly techniques for radioactive facilities;
- D.   Remote handling techniques needed for disassembly;
- E.   Radiation control techniques during dismantlement;
- F.   Waste management techniques for decommissioning;
- G.   Decommissioning systems engineering.

### Nuclear Power Engineering Test Center

Demonstration test of 1. dismantling techniques for reactor pressure vessels and 2. demolishing techniques for biological shield walls.


# NETHERLANDS


## NATIONAL POLICY


*Waste Management*:   Practice land-based storage of all categories of radioactive waste (LLW, ILW, HLW or spent fuel elements) for a period of several decades; decision as to disposal is to be deferred indefinitely. COVRA (Central Organisation for Radioactive Waste) has been organised to take responsibility for radioactive waste management for the Netherlands. No special policy has been developed for decommissioning.

*Decommissioning Strategy*:  It is generally envisaged that surplus nuclear facilities will have to be dismantled eventually but that this will not occur for several decades after the facility has been shut down.

*Financial Provisions for Decommissioning*:  No policy formulated; the owners of commercial power reactors may raise funds for decommissioning.

## REGULATIONS

No special regulations for decommissioning; such activities would be covered by the regulations governing public health and safety, as laid down in the Nuclear Energy Act.

## DEVELOPMENT AND DEMONSTRATION OF RELATED TECHNOLOGY

Special need: techniques for measuring residual radionuclide levels inside process equipment after decontamination, and techniques to qualify for unrestricted release.

# NORWAY

## NATIONAL POLICY

Four research reactors and one pilot reprocessing plant have been or are in operation in Norway, but no nuclear power is to be produced before the year 2000, hence little official attention has been paid to waste management or decommissioning.

## REGULATIONS

The operating licenses for the Norwegian research reactors, issued under the provisions of Act No. 28 of 12th May 1972 concerning nuclear energy activities, require that the owner of the facility prepare plans for decommissioning it and present the plans to the Nuclear Energy Safety Authority three years before the licensing permit is due to expire.

# SWEDEN

## NATIONAL POLICY

*Waste Management*:  Spent fuel and HLW canisters (from foreign reprocessor of Swedish spent fuel) are to be stored for 40 years before encapsulation and transfer to a geologic repository. Reactors wastes are to be disposed of underground, either at the reactor site or in a final repository that has storage chambers in bedrock, about 50 m below ground. Policy requires that the nuclear power plants be dismantled eventually, and storage chambers for decommissioning wastes will be excavated in the reactor waste repository in a later construction phase after the year 2000.

***Decommissioning Strategy***:   No decision has been taken between dismantling soon after final reactor shutdown and allowing a *cooling* period before dismantling. As the decommissioning of the first big nuclear power reactor in Sweden is still more than 20 years ahead, no detailed optimisation studies have been performed as yet.

***Financial Provisions for Decommissioning***:   The reactor owner pays an annual fee, based upon power production, into a fund managed by the state. When the reactor is decommissioned and dismantled, the reactor owner will be reimbursed for his costs from the fund. A similar provision is made for other back-end costs.

***Social Considerations***:   Nothing specific to decommissioning, but proposed nuclear legislation is subjected to extensive public review before presentation to the Parliament.

## REGULATIONS

The ***Act of Nuclear Activities*** of January 12, 1984 specifies that the holder of a license for nuclear activities is to:

1.   Take the measures necessary to maintain safety, safely handle and finally dispose of nuclear waste and decommission and dismantle in a safe manner plants in which the activity is no longer to be carried out (Section 10).
2.   Ensure that such research and development is conducted as is needed to fulfill the requirements set forth in Section 10 (Section 11).
3.   Prepare or have prepared a programme for the R&D work stipulated in Sections 10 and 11. The programme is to present a survey of all measures that may be necessary and also specify the measures that are intended to be taken within a period of at least six years. Beginning in 1986, the programme is to be submitted to the Government or designated authority every third year for examination and evaluation (Section 12).

The ***Act on Financing of Future Expenses for Spent Nuclear Fuel Etc.***, issued January 12, 1984, stipulates that the holder of a license to possess or operate a nuclear power reactor shall defray, inter alia, the costs for:

1.   the safe handling and final safe disposal of spent nuclear fuel from the reactor and radioactive waste deriving from it;
2.   the safe decommissioning and dismantling of the reactor installation;
3.   the required R&D.

Costs of these actions are to be estimated annually and an annual fee, adjusted each year, is to be paid to the State as long as the reactor is in operation.

Measures concerning the health and safety of workers and the public are prescribed in the Radiation Protection Act, while detailed regulations are provided by the National Institute for Radiation Protection.

## DEVELOPMENT AND DEMONSTRATION OF RELATED TECHNOLOGY

1.   Development of methods for system decontamination – Swedish nuclear power plants.
2.   Planning for the exchange of the steam generators at Ringhals 2 (to start in 1988).
3.   Start of disposal of reactor operational waste in the underground repository – 1988.

## DECOMMISSIONING COSTS

See Table 14 in Appendix 2.

# UNITED KINGDOM

## NATIONAL POLICY

*Waste Management*:  Reprocess spent fuel, vitrify the resulting HLW and store the HLW glass for at least 50 years; continue land disposal of LLW; establish new land disposal sites or repositories for LLW and ILW, including a deep repository for alpha-bearing wastes; replant existing power station sites with new units after obsolete plant has been taken out of service whenever such sites are suitable for redevelopment. (Generally, nuclear sites will remain in use as centres of generation for long after the present stations have been closed down, and obsolete reactor plants will, in many cases, be allowed to *cool* for many years before being dismantled).

*Decommissioning Strategy*:  The Central Electricity Generating Board (CEGB) has developed a preliminary concept for decommissioning Sizewell B:

1. *Stage 1* (commencing soon after shutdown) – remove all fuel; if shown to be advantageous, decontaminate the reactor primary loop, heat exchangers and other facilities; process and package sludges, resins and other radioactive wastes for offsite disposal.

2. *Stage 2* – dismantle the plant and buildings external to the reactor biological shield and remove waste materials. Completion of Stage 2 is expected about 10 years after reactor shutdown.

3. *Stage 3* – dismantle the reactor plant. Stage 3 might start immediately after Stage 2 is completed, or might be delayed several decades.

*Financial Provisions for Decommissioning*:  CEGB makes financial provision for the eventual costs of decommissioning in its accounts.

## REGULATIONS

No statutory regulations that relate specifically to decommissioning. It is anticipated that a nuclear plant operator will continue to be responsible for the safety of its operators and the general public until the Health and Safety Executive agrees that no danger exists from ionising radiations from anything on the site or any part thereof.

## DEVELOPMENT AND DEMONSTRATION OF RELATED TECHNOLOGY

1. Central Electricity Generating Board (CEGB) – formulation of a plan to decommission a typical Magnox power station.
2. United Kingdom Atomic Energy Authority – demonstration of equipment and procedures by decommissioning the Windscale Advanced Gas-Cooled Reactor.

## ESTIMATED DECOMMISSIONING COSTS

A Central Electricity Generating Board (CEGB) study assessed the total decommissioning costs for a Magnox station similar to Sizewell A. Assuming the use of today's technology and completion of the project through to final site clearance, the study concluded the project would cost between £270 million and £150 million at 1982 prices, depending upon the timing.

# UNITED STATES

## NATIONAL POLICY

*Waste Management*:   Disposal of LLW by shallow-land burial. Disposal of civilian HLW or spent fuel is defined by the Nuclear Waste Policy Act of 1982. This act establishes rules for siting, licensing, constructing and operating geological repositories for disposal of HLW. It also establishes a waste fund to be collected from the utilities and a timetable, leading to receipt of waste for disposal, beginning in 1998. The act calls for two repositories situated in different rock formations. All waste contaminated above 100 nCi TRU/g is stored retrievably for eventual disposal in a geological repository.

*Decommissioning Strategy*:   The NRC has proposed a comprehensive policy for decommissioning of nuclear facilities. When finally established, this policy is expected to be that all alternatives lead eventually to release of the site for unrestricted use (Stage 3) and termination of the license. Delays in reaching this condition, such as putting the plant in a safe storage (Stage 1) or entombment (Stage 2) condition, will be limited to situations where there is a compensating benefit, such as reduced occupational exposures due to radioactive decay or when there is an impediment such as lack of off-site spent fuel storage or waste disposal facilities. Planning for decommissioning will be specified at the licensing stage and updated near the end of plant operation. If a detailed cost estimate is not available, an estimate of at least $100 million (1984 dollars) will be used for a powerplant decommissioning. Methods for assuring availability of funds will be required. Some acceptable methods are: prepayment; external sinking fund (for utilities with single nuclear powerplants); surety method of insurance; an internal reserve fund. Currently, the role of the NRC in this area is to establish policy guidelines for decommissioning. If the NRC does not choose to take a more active role in this area, utilities will then be free to choose any decommissioning option that satisfies the guidelines. As a result, the option used will only require approval from the appropriate State regulatory commission. The same is true for the choice of a mechanism for recovering the funds.

The current decommissioning policies of 17 states were reviewed. Prompt dismantlement (Stage 3) is the preferred decommissioning option, followed by the safe storage method. Currently no State has chosen the entombment (Stage 2) option. It appears that most States prefer a prompt solution of the decommissioning problem. Further, it is interesting to note that most States will require that the land be restored to unrestricted or preconstruction uses.

*Financial Provisions for Decommissioning*:   The electric utility companies are required to establish a fund for decommissioning nuclear facilities. The Tax Reform Act of 1984 recognises the deductibility of utility contributions to a nuclear decommissioning trust fund. NRC requires proof that license applicants are financially able to safely remove reactors from operation.

All the regulatory bodies involved have considered decommissioning expenses to be a *cost of service*. A number of mechanisms have therefore been suggested to recover from ratepayers the funds used for decommissioning. One method, funding at commissioning, would commit the funds for decommissioning when a unit enters operation. Presumably, the funds would then be reinvested, and consequently only a fraction of the total cost would be prepaid. The prepaid funds, plus an appropriate return, would be recovered from ratepayers over the life of the plant. Alternatively, the funds could be accrued over the plant's operational lifetime. One method of doing this would be to collect revenues from ratepayers and place them in a *sinking* fund, which would be managed either by the utility (internal fund) or by a trustee (external fund). The *negative net salvage* method would then be used to accrue the funds over the life of the plant. Under this approach, the decommissioning expenses would be treated as *negative net salvage* for computing the capital cost for depreciation purposes. The costs would then be recovered by means of depreciation charges. This approach would treat decommissioning costs as just another capital cost; hence, the funds would remain *internal* to the utility. It is important to note that if the funds are *invested* at the same rate, then all three funds will yield the same annual charge to the ratepayers.

Eight states currently prefer the *sinking fund* mechanism for cost recovery from ratepayers. In a number of these states, legislation has already been enacted to create decommissioning committees to manage the funds, review plans, and adjust payment schedules. However, five other states have chosen to collect the necessary revenues by means of depreciation charges and to keep the decommissioning funds internal to the utilities.

*Social Considerations*:   The Decommissioning of a facility must comply with the National Environmental Policy Act (NEPA) of 1969. NEPA is the basic national charter for protection of the environment. It establishes policy, sets goals, and provides means for carrying out the policy. NEPA procedures ensure that environmental information is available to public officials and citizens before decisions are made and before actions are taken on such operations as the decommissioning of specific nuclear facilities.

# REGULATIONS

Federal government interests in the civilian nuclear power area are administered by the Department of Energy (DOE: R&D, uranium enrichment, waste disposal), the Nuclear Regulatory Commission (NRC: regulation and licensing) and the Environmental Protection Agency (EPA: environmental protection criteria). Commercial power generation, fuel fabrication, reprocessing and waste treatment activities are the responsibility of private industry. Fuel cycle and waste management R&D is conducted primarily by contractor organisations operating DOE's National Laboratories, repository projects, and the nuclear defence materials programme.

Although regulations covering all aspects of decommissioning utility powerplant and supporting fuel cycle facilities are not yet in place, the principal framework should be established within the next five years by the Nuclear Regulatory Commission (NRC) and the Environmental Protection Agency (EPA). The EPA is developing radiation protection criteria for the unrestricted release, after decontamination of land and facilities contaminated with radioactive materials. Some key issues to be dealt with are: whether the criteria will be broad guidance or more specific environmental standards; whether different criteria will be needed for several categories of facilities; how the level and longevity of the residual contamination will be addressed by the criteria; whether a time limit for reliance on institutional control is needed and what limit is appropriate; whether a level of residual radioactivity can be specified which is *below regulatory concern*.

NRC Regulatory Guide 1.86, Termination of Operating and Procedures Licenses for Nuclear Reactors describes methods for termination of an operating license, including the requirement that the dismantlement of the facility and disposal of the component parts not be inimical to the common defence and security or to the health and safety of the public.

Four alternatives of nuclear reactor facilities are considered acceptable in this guidance:

– Mothballing;
– In-Place Entombment;
– Removal of Radioactive Components and Dismantling;
– Conversion to a New Nuclear System or a Fossil Fuel System.

Prior to release of facilities and materials for unrestricted use, the licensee should make a comprehensive radiation survey establishing that contamination is within the following:

## ACCEPTABLE SURFACE CONTAMINATION LEVELS
dpm*/100 cm$^2$

| Nuclides | Average | Maximum | Removable |
|---|---|---|---|
| U-nat, U-235, U-238, and associated decay products | 5 000 | 15 000 | 1 000 |
| Transuranics, Ra-226, Ra-228, Th-230, Th-228, Pa-231, Ac-227, I-125, I-129 | 100 | 300 | 20 |
| Th-nat, Th-232, Sr-90, Ra-223, Ra-224, U-232, I-126, I-131, I-133 | 1 000 | 3 000 | 200 |
| Beta-gamma emitters (nuclides with decay modes other than alpha emission or spontaneous fission) except Sr-90 and others noted above. | 5 000 | 15 000 | 1 000 |

* dpm = disintegrations per minute.

Prior to release of the premises, a radiation survey of the surface and subsurface should be performed to assure that soil contamination is below guidelines for unrestricted use.

## DEVELOPMENT AND DEMONSTRATION OF RELATED TECHNOLOGY

Several DOE decommissioning projects including the continued clean-up work at TMI-2, the decommissioning at the West Valley Reprocessing Plant and the Shippingport Atomic Power Station will provide additional development and demonstration of decommissioning technology. A few utility power plants are in the process of being decommissioned and many utilities are preparing decommissioning plans and cost estimates for power plants.

# U.S. DOE DECOMMISSIONING PROJECTS

The decommissioning of Department of Energy (DOE) surplus facilities, which were previously used for research and development of nuclear power and the fuel cycle or for production of defence materials, will continue as part of the Surplus Facilities Management Programme (SFMP). There are currently about 350 facilities in the SFMP including reactor, fuel cycle support and materials production facilities, and waste disposal facilities. A comprehensive plan for decommissioning these facilities during the next 20 years at a cost exceeding $1.4 billion is being prepared, and is expected to be available during the next year. The fiscal year 1986 budget for this DOE programme is $56 million. Other DOE decommissioning activities will be conducted outside the SFMP, including decommissioning of small research reactors, accelerators, and nuclear materials research and utilisation facilities. A major DOE activity related to decommissioning of fuel cycle facilities is the stabilisation or relocation of inactive uranium mill tailings piles. Twenty-four piles will be stabilised or relocated and contaminated nearby properties will be cleaned up at a total cost of about $745 million through the early 1990s.

Another major DOE activity is the decontamination and decommissioning of the West Valley Reprocessing Plan. The DOE is removing high-level radioactive waste from the plant and the total project cost is estimated at $136 million through completion of waste solidification in 1990. The DOE is also participating in the Three Mile Island – 2 Recovery Programme along with the utilities to remove the fuel, and to decontaminate the facility. DOE's cost in the project is $189 million through completion in 1988.

## SFMP DECOMMISSIONING PROJECTS

### 1. Hanford Facilities

Several significant decommissioning projects are under way at Hanford, Washington.

The Strontium Semiworks complex, used for radioactive reprocessing has been maintained in a safe storage mode, requiring routine surveillance and maintenance since 1967. The complex will be decommissioned by a combination of dismantlement by explosives and entombment. The major process cell will be demolished to three metres above grade with explosives. These cell walls will be sheared off toward the interior with the rubble collecting in the bottom of the cell. The uncontaminated remaining structure will be knocked down with conventional wrecking equipment. An earthen barrier approximately three metres thick will be placed over the entombed structures and the area designated to low-level waste burial site.

The Hanford Production Reactors have been maintained in a safe storage mode, requiring routine surveillance and maintenance since 1965. In-situ decommissioning, (entombment) has been identified as the preferred decommissioning alternative, but the final selection of the facility disposition cannot be made until after the environmental review process. This alternative would involve leaving the reactor graphite block and associated heavy shielding walls in place. The perimeter portions of the reactor building would be demolished, leaving the reactor block intact on its foundation. Major voids in and around the reactor block would be filled and any openings would be sealed. Then the block would be covered over to a minimum depth of 5 metres using building rubble and natural gravel. The resulting mound would act as a long-term protective barrier to isolate the radioactive material from man and the environment.

### 2. Mound Facilities

Several facilities located at Mound Laboratories in Ohio managed large quantities of Pu-238 for fabrication of radioactive heat sources. These facilities will require extensive decontamination during decommissioning. One facility for decommissioning is the Special Metallurgical Building, which will be dismantled in a radiologically safe manner. The project includes the decontamination and decommissioning of this building, the associated facilities, and the surrounding area which were contamined with Pu-238 during operation.

Another Mound project is the decontamination of the Plutonium Processing Building, the Research Building and the Waste Transfer System. This project involves the removal of thousands of linear feet of gloveboxes; conveyor housings; and underground liquid waste lines. The estimated waste volume generated by the decommissioning project will be 380 000 cubic feet of waste contaminated with plutonium. Contamination in certain areas is being reduced to *As Low as Reasonably Achievable* (ALARA). The remaining contamination is permanently sealed so that the areas can be reused with minimal restrictions.

### 3. *Oak Ridge National Laboratories*

Numerous facilities located at Oak Ridge, Tennessee will also require extensive decontamination during decommissioning.

One major project, the Molten Salt Reactor Experiment (MSRE), was an experimental reactor which employed a molten-salt containing uranium-233 fuel and thorium in a circulated system. It was operated from June 1965 to December 1969 at a nominal full power level of 8.0 MW. It is a very highly radioactively contaminated system. Decommissioning activities will involve two major tasks:

1. permanent disposition of the fuel and flush salts, and
2. decontamination and decommissioning of the remaining reactor facilities.

Decommissioning planning for this project is in the preliminary planning stage. Detailed alternative assessments will begin in 1986. Initial work will focus on the proper disposal of the fuel and flush salt. This will involve development of containment techniques for salt disposal. A process for this must be developed and demonstrated. This task will be dependent on the determination of acceptable waste forms and packaging concepts for disposal. This fuel disposal task is expected to require at least 10 years. After processing of the fuel the MSRE facility may be decontamined and demolished, but no decision has yet been reached on the facility disposition.

### 4. *Shippingport Station Decommissioning Project*

The Shippingport Station is the first commercial nuclear power plant to undergo decommissioning in the United States. A primary objective of the Shippingport Station Decommissioning Project (SSDP) is to demonstrate project management and decommissioning techniques applicable to dismantling nuclear power plants. The dismantling of a plant of this size and complexity has not been undertaken to date.

The dismantling of the plant begins in September 1985. Asbestos will be removed from components and piping throughout the plant prior to removal of the components. Radioactive items will be removed first, followed by non-radioactive items. The reactor pressure vessel, with the reactor closure head and internals remaining in-place after defueling and with the neutron shield tank, will be removed in one piece and shipped by barge to the radioactive waste disposal area at Hanford, Washington. This package will be prepared by filling the neutron shielding tank with concrete, closing all reactor pressure vessel and closure head penetrations with welded shield plugs, installing a lifting beam and lifting skirt filled with concrete to shield the upper sides of the package, and anchoring the bottom end of the reactor pressure vessel to the neutron shield tank. Other large radioactive components such as the pressuriser, flash and blow-off tanks, steam generator heat exchangers, and reactor coolant pumps will also be sealed, externally decontaminated, and shipped as one piece on the barge to Hanford.

The reactor vessel package will be prepared and facilities for the lift and on-site transport of the package will be constructed in parallel with the general removal of systems and components. After components are removed, the remaining steel and concrete structures are to be surveyed and decontaminated as required and released for further dismantlement by standard demolition methods. The boiler and auxiliary chambers and enclosures are first to be cleared of all major components and radioactivity, except for the sealed steam generator heat exchangers and steam drums, before the roof of the enclosures are removed to gain access for efficient removal of the chamber steel. The heat exchangers and steam drums are to be removed as soon as adequate openings are made in the boiler chamber and the enclosure roof.

The SSDP project cost breakdown is as follows:

| | |
|---|---:|
| Conceptual Engineering | 540 |
| Engineering | 5 460 |
| Decommissioning Operations | 78 270 |
| Maintenance and surveillance | 3 250 |
| Contingency | 10 780 |
| Total project cost ($ in thousands) | 98 300 |

# OTHER DOE PROJECTS

## 1.    *West Valley Reprocessing Plant*

The Reprocessing Plant and associated facilities at the Western New York Nuclear Service Centre (WNYNSC) will be decommissioned in conjunction with solidification of the high-level radioactive waste (HLW) presently stored in underground tanks at the WNYNSC.

Approximately 2 270 cubic metres of HLW in liquid and sludge form is stored in two specially designed and constructed underground tank storage systems adjacent to the Processing Plant building. This HLW will be processed into a solid vitrified form of borosilicate glass encapsulated in stainless steel canisters. New glass melter equipment is being assembled on the WNYNSC site to provide the HLW Vitrification Process. The shielded cells in the Reprocessing Plant building are being decontaminated and dismantled in order to install components and subsystems for the HLW Vitrification Process, the reduction and immobilisation of low-level radioactive waste (LLW) and transuranic wastes (TRU), and for the interim storage of the vitrified HLW canisters as well as the packaged TRU and LLW.

The post-vitrification decommissioning plan for the WNYNSC has not been finalised, but will include *removal of radioactive components from* the Processing Plant building and the glass melter building. The current reference is that the building will be decontaminated to a radiation level suitable for uncontrolled use. The underground HLW storage tanks will be emptied of their liquid and sludge HLW, and entombed in accordance with applicable regulations at that time. The major post-vitrification decommissioning activities are scheduled to begin in 1991 and to be completed in 1998.

## 2.    *Three Mile Island-2 Recovery Programme*

The programme at Three Mile Island Unit 2 involves the removal of irradiated fuel debris and radioactive wastes as well as nuclear safety research on the cause and course of the accident. Nearly every system and component in the Reactor Containment Building and support facilities was affected in some way by the accident and requires recovery activities. A decontamination facility utilising electropolishing, freon spray cleaning, freon hose and cable cleaning, vibratory finishing, and both ultrasonic cleaning for tools and small pieces of equipment was constructed and is being operated.

Contaminated water has been removed from the auxiliary building and has been processed. The auxiliary building walls have been cleaned and painted to affix the contamination to the walls. The equipment is being decontaminated. Contaminated water has been removed from the reactor building and processed. Shielding has been installed to reduce exposure; and water spraying, scabbling and painting of building floors has been performed to reduce working exposure levels. Examination, repair and replacement of parts has been performed to refurbish the polar crane which will be used in subsequent operations. The reactor head has been shielded and removed from the reactor vessel and placed on the storage stand to allow for core inspection and clean-up.

In future years, remote processing and removal of highly contaminated resins from the purification demineralisers will be conducted. Resins contaminated with fuel and cesium will be eluted to remove high cesium activity and be sluiced and packaged for shipment and ultimate disposition. Several robots will be used for characterising high radiation areas, performing special tasks such as decontamination and will assist in reactor defuelling. Removal of the fuel debris from the reactor will be accomplished with special tools, a robotic arm and debris canisters.

### *U.S. Utility Decommissioning Projects*

In addition to DOE activities, utility power plants at Humboldt Bay and Dresden 1 are being decommissioned. As part of meeting regulatory requirements and the inclusion of decommissioning funds in the rate base, utilities are preparing decommissioning plans and cost estimates for each plant. The following table shows estimated decommissioning costs for selected U.S. nuclear power plans. Studies are also under way sponsored by utility organisations for extending the lifetimes of operating nuclear power plants.

# ESTIMATED DECOMMISSIONING COSTS FOR SELECTED U.S. NUCLEAR POWER PLANTS

| State of Juridiction | Plant, Unit | Reactor[a] Type | Reactor[a] Capacity Net (MWe) | Utility's Estimated Date of Decommissioning | Utility's Estimated Cost Total (million 1983 dollars) | Utility's Estimated Cost Per Kilowatt (1983 dollars) |
|---|---|---|---|---|---|---|
| Alabama | Farley 1 | PWR | 829 | 2012 | 40.96 | 49.41 |
| | Farley 2 | PWR | 829 | 2012 | 40.02 | 48.28 |
| Arkansas | Arkansas Nuclear One 1 | PWR | 850 | 2009-2014 | 29.43 | 34.62 |
| | Arkansas Nuclear One 2 | PWR | 912 | 2011-2016 | 27.04 | 29.65 |
| California | San Onofre 1 | PWR | 436 | 1997 | 110.25 | 252.86 |
| | San Onofre 2 | PWR | 1 070 | 2015 | 120.99 | 113.07 |
| | San Onofre 3 | PWR | 1 080 | 2015 | 137.02 | 126.87 |
| | Diablo Canyon 1 | PWR | 1 084 | 2015 | 104.81 | 95.72 |
| | Diablo Canyon 2 | PWR | 1 084 | 2015 | 104.81 | 95.72 |
| | Humboldt Bay 3 | BWR | 65 | 1984-2014[b] | 52.15 | 802.24 |
| Connecticut | Millstone 1 | BWR | 660 | 2006 | 93.80 | 142.12 |
| | Millstone 2 | PWR | 870 | 2010 | 94.70 | 108.85 |
| Florida | Turkey Point 3 | PWR | 693 | 2007 | 80.28 | 115.85 |
| | Turkey Point 4 | PWR | 693 | 2007 | 58.57 | 84.52 |
| | St. Lucie 1 | PWR | 830 | 2010 | 86.62 | 104.36 |
| | St. Lucie 2 | PWR | 804 | 2023 | 74.52 | 92.64 |
| | Crystal River 3 | PWR | 825 | 2008 | 84.44 | 102.35 |
| Iowa | Quad Cities 1[c] | BWR | 789 | 2007 | NA | NA |
| | Quad Cities 2[c] | BWR | 789 | 2007 | NA | NA |
| | Duane Arnold 1 | BWR | 538 | 2004 | NA | NA |
| | Cooper 1[d] | BWR | 778 | 2004 | NA | NA |
| Kansas | Wolf Creek | PWR | 1 150 | NA | NA | NA |
| Maryland | Calvert Cliffs 1 | PWR | 845 | 2009 | 46.88 | 55.48 |
| | Calvert Cliffs 2 | PWR | 845 | 2009 | 46.88 | 55.48 |
| Michigan | Enrico Fermi 1 | FBR | 61 | NA | 12.44 | 203.91 |
| | Enrico Fermi 2 | BWR | 1 093 | NA | 121.34 | 111.02 |
| | Big Rock 1 | PWR | 72 | 2000 | 47.17 | 655.13 |
| | Palisades 1 | PWR | 805 | 2007 | 97.06 | 120.57 |
| New Hampshire | Seabrook 1 | PWR | 1 198 | NA | 137.34 | 114.64 |
| | Seabrook 2 | PWR | 1 198 | NA | 137.34 | 114.64 |
| New Jersey | Salem 1[e] | PWR | 1 090 | 2008 | 43.00 | 39.45 |
| New York | Ginna 1 | PWR | 470 | NA | 52.92 | 112.60 |
| | Indian Point 2 | PWR | 873 | NA | 72.73 | 86.31 |
| | Nine Mile Point 1 | BWR | 620 | NA | 69.61 | 112.28 |
| North Carolina | Brunswick 1 | BWR | 821 | 2000 | 61.59 | 75.02 |
| | Brunswick 2 | BWR | 821 | 2000 | 81.59 | 99.87 |
| | McGuire 1 | PWR | 1 180 | 2010 | 55.11 | 46.70 |
| | McGuire 2 | PWR | 1 180 | 2013 | 55.11 | 46.70 |
| Pennsylvania | Beaver Valley | PWR | 835 | 2010 | 64.60 | 77.36 |
| | Peach Bottom 2[e] | BWR | 1 065 | 2008 | 58.10 | 54.55 |
| | Peach Bottom 3[e] | BWR | 1 065 | 2008 | 58.10 | 54.55 |
| | Susquehanna 1 | BWR | 1 065 | 2023 | 86.50 | 81.22 |
| | Three Mile Island 1[e] | PWR | 819 | 2011 | 37.70 | 46.03 |
| South Carolina | Oconee 1 | PWR | 887 | 2003 | 60.07 | 67.72 |
| | Oconee 2 | PWR | 887 | 2003 | 60.07 | 67.72 |
| | Oconee 3 | PWR | 887 | 2003 | 60.07 | 67.72 |
| | Robinson 2 | PWR | 700 | 1997 | 53.63 | 76.61 |
| | V.C. Summer 1 | PWR | 900 | 2012 | 50.25 | 55.96 |

*(Table cont'd next page)*

## ESTIMATED DECOMMISSIONING COSTS FOR SELECTED U.S. NUCLEAR POWER PLANTS

| State of Juridiction | Plant, Unit | Reactor[a] | | Utility's Estimated Date of Decommissioning | Utility's Estimated Cost | |
| --- | --- | --- | --- | --- | --- | --- |
| | | Type | Capacity Net (MWe) | | Total (million 1983 dollars) | Per Kilowatt (1983 dollars) |
| Texas | Comanche Peak 1 | PWR | 1 150 | 2115 | 104.00 | 90.43 |
| | Comanche Peak 2 | PWR | 1 150 | 2116 | 104.00 | 90.43 |
| | Palo Verde 1 | PWR | 1 304 | 2114 | 83.20 | 63.80 |
| | Palo Verde 2 | PWR | 1 304 | 2115 | 83.20 | 63.80 |
| | Palo Verde 3 | PWR | 1 304 | 2116 | 83.20 | 63.80 |
| | River Bend 1 | BWR | 934 | 2116 | NA | NA |
| | South Texas 1 | PWR | 1 250 | 2117 | NA | NA |
| | South Texas 2 | PWR | 1 250 | 2119 | NA | NA |
| Virginia | North Anna 1 | PWR | 907 | 2010 | 118.38[f] | 65.26 |
| | North Anna 2 | PWR | 907 | 2010 | | 65.26 |
| | Salem 1 | PWR | 1 090 | 2008 | NA | NA |
| | Salem 2 | PWR | 1 115 | 2008 | NA | NA |
| | Peach Bottom 2 | BWR | 1 065 | 2008 | NA | NA |
| | Peach Bottom 3 | BWR | 1 065 | 2008 | NA | NA |
| | Surry 1 | PWR | 788 | 2003 | 112.42[f] | 71.33 |
| | Surry 2 | PWR | 788 | 2003 | | 71.33 |
| Wisconsin | Kewaunee | PWR | 535 | 2008 | 71.35 | 133.37 |
| | Point Beach 1 | PWR | 497 | 2007 | 82.46 | 105.36 |
| | Point Beach 2 | PWR | 497 | 2008 | 82.46 | 105.36 |

a) Reactor type: BWR, boiling-water reactor; FBR, fast breeder reactor; PWR, pressurized-water reactor.
b) Safe storage to begin in 1984, dismantlement in 2014.
c) Plant is located in Illinois, but one-fourth ownership interest is in Iowa.
d) Plant is located in Nebraska, but one-half ownership interest is in Iowa.
e) Because of joint ownership or interstate electricity sales, the reported decommissioning cost may not reflect 100 per cent of the estimated cost.
f) Combined cost for both units at the site.
NA = not available.
*Source:* R.A. Cantor, *Nuclear Power Decommissioning: Analysis of Regulatory Environments,* TM-244/85 (Oak Ridge National Laboratory, Oak Ridge, Tennessee, 1985).

# COMMISSION OF THE EUROPEAN COMMUNITIES

## DEVELOPMENT AND DEMONSTRATION OF RELATED TECHNOLOGY

Research and development projects concerning the following subjects are being carried out:

1. Long-term integrity of buildings and systems;
2. Decontamination for decommissioning purposes;
3. Dismantling techniques;
4. Treatment of specific waste material: steel, concrete and graphite;
5. Large containers for radioactive waste produced in the dismantling of nuclear installations;
6. Estimation of the quantities of radioactive wastes arising from the decommissioning of nuclear installations in the Community;
7. Influence of installation design features on decommissioning.

Moreover, new decommissioning techniques are being tested under real conditions, within the framework of large-scale decommissioning operations.

# OECD SALES AGENTS
# DÉPOSITAIRES DES PUBLICATIONS DE L'OCDE

**ARGENTINA - ARGENTINE**
Carlos Hirsch S.R.L.,
Florida 165, 4º Piso,
(Galeria Guemes) 1333 Buenos Aires
Tel. 33.1787.2391 y 30.7122

**AUSTRALIA-AUSTRALIE**
D.A. Book (Aust.) Pty. Ltd.
11-13 Station Street (P.O. Box 163)
Mitcham, Vic. 3132          Tel. (03) 873 4411

**AUSTRIA - AUTRICHE**
OECD Publications and Information Centre,
4 Simrockstrasse,
5300 Bonn (Germany)      Tel. (0228) 21.60.45
Local Agent:
Gerold & Co., Graben 31, Wien 1   Tel. 52.22.35

**BELGIUM - BELGIQUE**
Jean de Lannoy, Service Publications OCDE,
avenue du Roi 202
B-1060 Bruxelles               Tel. 02/538.51.69

**CANADA**
Renouf Publishing Company Limited/
Éditions Renouf Limitée Head Office/
Siège social – Store/Magasin :
61, rue Sparks Street,
Ottawa, Ontario KIP 5A6
          Tel. (613)238-8985. 1-800-267-4164
Store/Magasin : 211, rue Yonge Street,
Toronto, Ontario M5B 1M4.
          Tel. (416)363-3171
Regional Sales Office/
Bureau des Ventes régional :
7575 Trans-Canada Hwy., Suite 305,
Saint-Laurent, Quebec H4T 1V6
          Tel. (514)335-9274

**DENMARK - DANEMARK**
Munksgaard Export and Subscription Service
35, Nørre Søgade, DK-1370 København K
          Tel. +45.1.12.85.70

**FINLAND - FINLANDE**
Akateeminen Kirjakauppa,
Keskuskatu 1, 00100 Helsinki 10   Tel. 0.12141

**FRANCE**
OCDE/OECD
Mail Orders/Commandes par correspondance :
2, rue André-Pascal,
75775 Paris Cedex 16
          Tel. (1) 45.24.82.00
Bookshop/Librairie : 33, rue Octave-Feuillet
75016 Paris
          Tel. (1) 45.24.81.67 et/ou (1) 45.24.81.81
Principal correspondant :
Librairie de l'Université,
12α, rue Nazareth,
13602 Aix-en-Provence      Tel. 42.26.18.08

**GERMANY - ALLEMAGNE**
OECD Publications and Information Centre,
4 Simrockstrasse,
5300 Bonn              Tel. (0228) 21.60.45

**GREECE - GRÈCE**
Librairie Kauffmann,
28, rue du Stade, 105 64 Athens   Tel. 322.21.60

**HONG KONG**
Government Information Services,
Publications (Sales) Office,
Beaconsfield House, 4/F.,
Queen's Road Central

**ICELAND - ISLANDE**
Snæbjörn Jónsson & Co., h.f.,
Hafnarstræti 4 & 9,
P.O.B. 1131 – Reykjavik
          Tel. 13133/14281/11936

**INDIA - INDE**
Oxford Book and Stationery Co.,
Scindia House, New Delhi I      Tel. 45896
17 Park St., Calcutta 700016      Tel. 240832

**INDONESIA - INDONESIE**
Pdii-Lipi, P.O. Box 3065/JKT.Jakarta
          Tel. 583467

**ITALY - ITALIE**
Libreria Commissionaria Sansoni,
Via Lamarmora 45, 50121 Firenze
          Tel. 579751/584468
Via Bartolini 29, 20155 Milano    Tel. 365083
Sub-depositari :
Ugo Tassi, Via A. Farnese 28,
00192 Roma              Tel. 310590
Editrice e Libreria Herder,
Piazza Montecitorio 120, 00186 Roma
          Tel. 6794628
Agenzia Libraria Pegaso,
Via de Romita 5, 70121 Bari
          Tel. 540.105/540.195
Agenzia Libraria Pegaso, Via S.Anna dei
Lombardi 16, 80134 Napoli.      Tel. 314180
Libreria Hœpli,
Via Hœpli 5, 20121 Milano       Tel. 865446
Libreria Scientifica
Dott. Lucio de Biasio "Aeiou"
Via Meravigli 16, 20123 Milano   Tel. 807679
Libreria Zanichelli, Piazza Galvani 1/A,
40124 Bologna              Tel. 237389
Libreria Lattes,
Via Garibaldi 3, 10122 Torino    Tel. 519274
La diffusione delle edizioni OCSE è inoltre
assicurata dalle migliori librerie nelle città più
importanti.

**JAPAN - JAPON**
OECD Publications and Information Centre,
Landic Akasaka Bldg., 2-3-4 Akasaka,
Minato-ku, Tokyo 107      Tel. 586.2016

**KOREA - CORÉE**
Pan Korea Book Corporation
P.O.Box No. 101 Kwangwhamun, Seoul
          Tel. 72.7369

**LEBANON - LIBAN**
Documenta Scientifica/Redico,
Edison Building, Bliss St.,
P.O.B. 5641, Beirut      Tel. 354429-344425

**MALAYSIA - MALAISIE**
University of Malaya Co-operative Bookshop
Ltd.,
P.O.Box 1127, Jalan Pantai Baru,
Kuala Lumpur          Tel. 577701/577072

**NETHERLANDS - PAYS-BAS**
Staatsuitgeverij
Chr. Plantijnstraat, 2 Postbus 20014
2500 EA S-Gravenhage      Tel. 070-789911
Voor bestellingen:          Tel. 070-789880

**NEW ZEALAND - NOUVELLE-ZÉLANDE**
Government Printing Office Bookshops:
Auckland: Retail Bookshop, 25 Rutland Street,
Mail Orders, 85 Beach Road
Private Bag C.P.O.
Hamilton: Retail: Ward Street,
Mail Orders, P.O. Box 857
Wellington: Retail, Mulgrave Street, (Head
Office)
Cubacade World Trade Centre,
Mail Orders, Private Bag
Christchurch: Retail, 159 Hereford Street,
Mail Orders, Private Bag
Dunedin: Retail, Princes Street,
Mail Orders, P.O. Box 1104

**NORWAY - NORVÈGE**
Tanum-Karl Johan
Karl Johans gate 43, Oslo 1
PB 1177 Sentrum, 0107 Oslo 1Tel. (02) 42.93.10

**PAKISTAN**
Mirza Book Agency
65 Shahrah Quaid-E-Azam, Lahore 3 Tel. 66839

**PORTUGAL**
Livraria Portugal,
Rua do Carmo 70-74, 1117 Lisboa Codex.
          Tel. 360582/3

**SINGAPORE - SINGAPOUR**
Information Publications Pte Ltd
Pei-Fu Industrial Building,
24 New Industrial Road No. 02-06
Singapore 1953          Tel. 2831786, 2831798

**SPAIN - ESPAGNE**
Mundi-Prensa Libros, S.A.,
Castelló 37, Apartado 1223, Madrid-28001
          Tel. 431.33.99
Libreria Bosch, Ronda Universidad 11,
Barcelona 7          Tel. 317.53.08/317.53.58

**SWEDEN - SUÈDE**
AB CE Fritzes Kungl. Hovbokhandel,
Box 16356, S 103 27 STH,
Regeringsgatan 12,
DS Stockholm          Tel. (08) 23.89.00
Subscription Agency/Abonnements:
Wennergren-Williams AB,
Box 30004, S104 25 Stockholm. Tel. 08/54.12.00

**SWITZERLAND - SUISSE**
OECD Publications and Information Centre,
4 Simrockstrasse,
5300 Bonn (Germany)      Tel. (0228) 21.60.45
Local Agent:
Librairie Payot,
6 rue Grenus, 1211 Genève 11
          Tel. (022) 31.89.50

**TAIWAN - FORMOSE**
Good Faith Worldwide Int'l Co., Ltd.
9th floor, No. 118, Sec.2
Chung Hsiao E. Road
Taipei          Tel. 391.7396/391.7397

**THAILAND - THAILANDE**
Suksit Siam Co., Ltd.,
1715 Rama IV Rd.,
Samyam Bangkok 5          Tel. 2511630

**TURKEY - TURQUIE**
Kültur Yayinlari Is-Türk Ltd. Sti.
Atatürk Bulvari No: 191/Kat. 21
Kavaklidere/Ankara          Tel. 25.07.60
Dolmabahce Cad. No: 29
Besiktas/Istanbul          Tel. 160.71.88

**UNITED KINGDOM - ROYAUME UNI**
H.M. Stationery Office,
Postal orders only:
P.O.B. 276, London SW8 5DT
Telephone orders: (01) 622.3316, or
Personal callers:
49 High Holborn, London WC1V 6HB
Branches at: Belfast, Birmingham,
Bristol, Edinburgh, Manchester

**UNITED STATES - ÉTATS-UNIS**
OECD Publications and Information Centre,
Suite 1207, 1750 Pennsylvania Ave., N.W.,
Washington, D.C. 20006 - 4582
          Tel. (202) 724.1857

**VENEZUELA**
Libreria del Este,
Avda F. Miranda 52, Aptdo. 60337,
Edificio Galipan, Caracas 106
          Tel. 32.23.01/33.26.04/31.58.38

**YUGOSLAVIA - YOUGOSLAVIE**
Jugoslovenska Knjiga, Knez Mihajlova 2,
P.O.B. 36, Beograd          Tel. 621.992

Orders and inquiries from countries where Sales
Agents have not yet been appointed should be sent
to:
OECD, Publications Service, Sales and
Distribution Division, 2, rue André-Pascal, 75775
PARIS CEDEX 16.

Les commandes provenant de pays où l'OCDE n'a
pas encore désigné de dépositaire peuvent être
adressées à :
OCDE, Service des Publications. Division des
Ventes et Distribution. 2. rue André-Pascal. 75775
PARIS CEDEX 16.

70024-10-1986

OECD PUBLICATIONS, 2, rue André-Pascal, 75775 PARIS CEDEX 16 - No. 43643 1986
PRINTED IN FRANCE
(66 86 08 1) ISBN 92-64-12894-8